Wireless Communications and Technology

Wireless Communications and Technology

Edited by
Trevor Miller

Larsen & Keller
www.larsen-keller.com

Wireless Communications and Technology
Edited by Trevor Miller
ISBN: 978-1-63549-299-6 (Hardback)

Larsen & Keller
Published by Larsen and Keller Education,
5 Penn Plaza,
19th Floor,
New York, NY 10001, USA

Cataloging-in-Publication Data

Wireless communications and technology / edited by Trevor Miller.
p. cm.
Includes bibliographical references and index.
ISBN 978-1-63549-299-6
1. Wireless communication systems. 2. Telecommunication.
3. Radio frequency. 4. Radio frequency identification
systems. I. Miller, Trevor.
TK5103.2 .W57 2017
621.384--dc23

Printed and bound in China.

For more information regarding Larsen and Keller Education and its products, please visit the publisher's website www.larsen-keller.com

Table of Contents

Preface

This book is compiled in such a manner that it will provide in-depth knowledge about the theory and practice of wireless communications and technology. It describes in detail the various concepts and technologies used in this subject. Wireless communication allows transfer of data in the form of text, voice and image between two points which are not connected via wires. Wireless communication technology is used in mobile and portable applications, wireless networks, and personal digital assistants (PDAs), etc. Most of the topics introduced in this text cover new techniques and applications of the subject. Different approaches, evaluations and methodologies on the subject have been included in it. For all those who are interested in wireless communications and technology, this textbook can prove to be an essential guide.

A foreword of all Chapters of the book is provided below:

Chapter 1 - Wireless communication is the information that has been transferred from one point to another without the use of an electrical conductor. Radio is the most common technology that is used in contemporary times. The chapter will provide an integrated understanding of wireless communication; **Chapter 2** - The networks mentioned in this chapter are wireless network, wireless mesh network, personal area network, space network, comparison of wireless data standard and multipath propagation. Wireless networks use wireless connections; they are mostly used in radio communications. The most common examples of wireless networks include cell phone networks, wireless sensor networks and satellite communication networks. This text elucidates the networks related to wireless communication; **Chapter 3** - This section is an overview of the subject matter incorporating all the major aspects of a radio. Some of the topics discussed in this text are radio propagation, radio frequency, skywave and radio broadcasting. The pattern in which radio waves travel is known as radio propagation whereas radio frequency is the frequency that is used for communication and lies in between 3 kHZ to 300 GHz. The topics discussed herein are of great importance to broaden the existing knowledge on radio; **Chapter 4** - To understand radio frequency identification, it is very important to understand real-time locating system, microchip implant and transponder timing. Real-time locating systems are used in tracking objects and people in real-time. This section elucidates the crucial aspects of radio frequency identification; **Chapter 5** - Radio navigation is an application that is used to trace an object on Earth whereas radars are used to determine the position and range of an object. Countries use radars to detect unknown aircrafts, ships, spacecrafts and guided missiles. The other wireless technologies and applications explained in this chapter are Wi-Fi, hotspot and Li-Fi; **Chapter 6** - Light is used to carry information from a distance and this form of communication is known as optical communication. One of the forms of optical communication is optical wireless communications. The aspects elucidated in this section are of vital importance, and provide a better understanding of optical wireless communications; **Chapter 7** - Wireless security is very important in today's times. It prevents unapproved access to computers that use wireless networks. A brief explanation of intrusion detection system, RADIUS and Wi-Fi protected access is also provided, which provides a better understanding of wireless security.

I would like to thank the entire editorial team who made sincere efforts for this book and my family who supported me in my efforts of working on this book. I take this opportunity to thank all those who have been a guiding force throughout my life.

Editor

1

Introduction to Wireless Communication

Wireless communication is the information that has been transferred from one point to another without the use of an electrical conductor. Radio is the most common technology that is used in contemporary times. The chapter will provide an integrated understanding of wireless communication.

Wireless

Wireless communication is the transfer of information or power between two or more points that are not connected by an electrical conductor. The most common wireless technologies use radio. With radio waves distances can be short, such as a few meters for television or as far as thousands or even millions of kilometers for deep-space radio communications. It encompasses various types of fixed, mobile, and portable applications, including two-way radios, cellular telephones, personal digital assistants (PDAs), and wireless networking. Other examples of applications of radio *wireless technology* include GPS units, garage door openers, wireless computer mice, keyboards and headsets, headphones, radio receivers, satellite television, broadcast television and cordless telephones.

Somewhat less common methods of achieving wireless communications include the use of other electromagnetic wireless technologies, such as light, magnetic, or electric fields or the use of sound. The term *wireless* has been used twice in communications history, with slightly different meaning. It was initially used from about 1890 for the first radio transmitting and receiving technology, as in *wireless telegraphy*, until the new word *radio* replaced it around 1920. The term was revived in the 1980s and 1990s mainly to distinguish digital devices that communicate without wires, such as the examples listed in the previous paragraph, from those that require wires or cables. This is its primary usage in the 2000s. LTE, LTE-Advanced, Wi-Fi and Bluetooth are common modern wireless technologies used in the 2000s.

Wireless operations permit services, such as long-range communications, that are impossible or impractical to implement with the use of wires. The term is commonly used in the telecommunications industry to refer to telecommunications systems (e.g. radio transmitters and receivers, remote controls, etc.) which use some form of energy (e.g. radio waves, acoustic energy,) to transfer information without the use of wires. Information is transferred in this manner over both short and long distances.

Photophone

The world's first wireless telephone conversation occurred in 1880, when Alexander Graham Bell and Charles Sumner Tainter invented and patented the photophone, a telephone that conducted audio conversations wirelessly over modulated light beams (which are narrow projections of electromagnetic waves). In that distant era, when utilities did not yet exist to provide electricity and lasers had not even been imagined in science fiction, there were no practical applications for their invention, which was highly limited by the availability of both sunlight and good weather. Similar to free-space optical communication, the photophone also required a clear line of sight between its transmitter and its receiver. It would be several decades before the photophone's principles found their first practical applications in military communications and later in fiber-optic communications.

Bell and Tainter's photophone, of 1880.

Early Usage

David E. Hughes transmitted radio signals over a few hundred yards using a clockwork keyed transmitter in 1878. As this was before Maxwell's work was understood, Hughes' contemporaries dismissed his achievement as mere "Induction." In 1885, Thomas Edison used a vibrator magnet for induction transmission. In 1888, Edison deployed a system of signaling on the Lehigh Valley Railroad. In 1891, Edison obtained the wireless patent for this method using inductance (U.S. Patent 465,971).

In 1888, Heinrich Hertz demonstrated the existence of electromagnetic waves, the underlying basis of most wireless technology. The theory of electromagnetic waves was predicted from the research of James Clerk Maxwell and Michael Faraday. Hertz demonstrated that electromagnetic waves traveled through space in straight lines, could be transmitted, and could be received by an experimental apparatus. Hertz did not follow up on the experiments. Jagadish Chandra Bose around this time developed an early wireless detection device and helped increase the knowledge of millimeter-length electromagnetic waves. Later inventors implemented practical applications of wireless radio communication and radio remote control technology.

Radio

The term "wireless" came into public use to refer to a radio receiver or transceiver (a dual purpose receiver and transmitter device), establishing its use in the field of wireless telegraphy early on; now the term is used to describe modern wireless connections such as in cellular networks and wireless broadband Internet. It is also used in a general sense to refer to any operation that is implemented without the use of wires, such as "wireless remote control" or "wireless energy transfer", regardless of the specific technology (e.g. radio, infrared, ultrasonic) used. Guglielmo Marconi and Karl Ferdinand Braun were awarded the 1909 Nobel Prize for Physics for their contribution to wireless telegraphy.

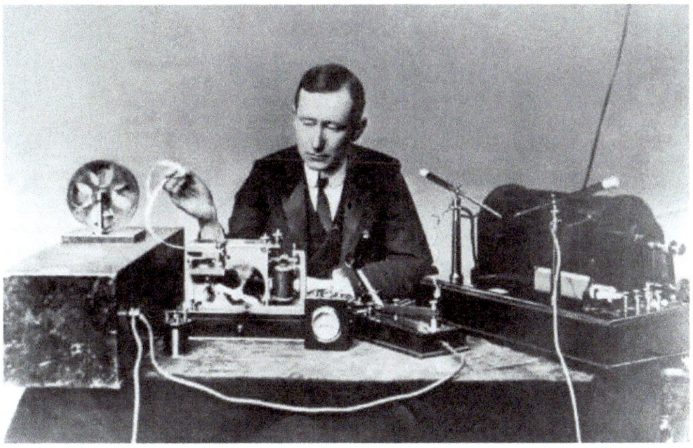

Marconi transmitting the first radio signal across the Atlantic.

Modes

Wireless communications can be via:

Radio

radio communication, microwave communication, for example long-range line-of-sight via highly directional antennas, or short-range communication.

Free-space Optical

An 8-beam free space optics laser link, rated for 1 Gbit/s at a distance of approximately 2 km. The receptor is the large disc in the middle, the transmitters the smaller ones. To the top and right corner a monocular for assisting the alignment of the two heads.

Free-space optical communication (FSO) is an optical communication technology that uses light propagating in free space to transmit wirelessly data for telecommunications or computer networking. "Free space" means the light beams travel through the open air or outer space. This contrasts with other communication technologies that use light beams traveling through transmission lines such as optical fiber or dielectric "light pipes".

The technology is useful where physical connections are impractical due to high costs or other considerations. For example, free space optical links are used in cities between office buildings which are not wired for networking, where the cost of running cable through the building and under the street would be prohibitive. Another widely used example is consumer IR devices such as remote controls and IrDA (Infrared Data Association) networking, which is used as an alternative to WiFi networking to allow laptops, PDAs, printers, and digital cameras to exchange data.

Sonic

Sonic, especially ultrasonic short range communication involves the transmission and reception of sound.

Electromagnetic Induction

Electromagnetic induction short range communication and power. This has been used in biomedical situations such as pacemakers, as well as for short-range Rfid tags.

Services

Common examples of wireless equipment include:

- Infrared and ultrasonic remote control devices
- Professional LMR (Land Mobile Radio) and SMR (Specialized Mobile Radio) typically used by business, industrial and Public Safety entities.
- Consumer Two-way radio including FRS Family Radio Service, GMRS (General Mobile Radio Service) and Citizens band ("CB") radios.
- The Amateur Radio Service (Ham radio).
- Consumer and professional Marine VHF radios.
- Airband and radio navigation equipment used by aviators and air traffic control
- Cellular telephones and pagers: provide connectivity for portable and mobile applications, both personal and business.
- Global Positioning System (GPS): allows drivers of cars and trucks, captains of boats and ships, and pilots of aircraft to ascertain their location anywhere on earth.
- Cordless computer peripherals: the cordless mouse is a common example; wireless headphones, keyboards, and printers can also be linked to a computer via wireless using technology such as Wireless USB or Bluetooth
- Cordless telephone sets: these are limited-range devices.

- Satellite television: Is broadcast from satellites in geostationary orbit. Typical services use direct broadcast satellite to provide multiple television channels to viewers.

Computers

- Wi-Fi
- Cordless computer peripherals:
    - mouse
    - headphones,
    - keyboards,
    - printers,
    - USB and,
    - Bluetooth
- Wireless networking
    - To span a distance beyond the capabilities of typical cabling,
    - To provide a backup communications link in case of normal network failure,
    - To link portable or temporary workstations,
    - To overcome situations where normal cabling is difficult or financially impractical, or
    - To remotely connect mobile users or networks.

Developers need to consider some parameters involving Wireless RF technology for better developing wireless networks:

- Sub-GHz versus 2.4 GHz frequency trends
- Operating range and battery life
- Sensitivity and data rate
- Network topology and node intelligence

Applications may involve point-to point communication, point-to-multipoint communication, broadcasting, cellular networks and other wireless networks, Wi-Fi technology.

Cordless

The term "wireless" should not be confused with the term "cordless", which is generally used to refer to powered electrical or electronic devices that are able to operate from a portable power source (e.g., a battery pack) without any cable or cord to limit the mobility of the cordless device through a connection to the mains power supply. Some cordless devices, such as cordless telephones, are also wireless in the sense that information is transferred from the cordless telephone to the phone's base unit via some wireless communications link. This has caused some disparity in the usage of the term "cordless", for example in Digital Enhanced Cordless Telecommunications.

Electromagnetic Spectrum

Light, colors, AM and FM radio, and electronic devices make use of the electromagnetic spectrum. The frequencies of the radio spectrum that are available for use for communication are treated as a public resource and are regulated by national organizations such as the Federal Communications Commission in the USA, or Ofcom in the United Kingdom. This determines which frequency ranges can be used for what purpose and by whom. In the absence of such control or alternative arrangements such as a privatized electromagnetic spectrum, chaos might result if, for example, airlines did not have specific frequencies to work under and an amateur radio operator were interfering with the pilot's ability to land an aircraft. Wireless communication spans the spectrum from 9 kHz to 300 GHz.

Applications

Mobile Telephones

One of the best-known examples of wireless technology is the mobile phone, also known as a cellular phone, with more than 4.6 billion mobile cellular subscriptions worldwide as of the end of 2010. These wireless phones use radio waves from signal-transmission towers to enable their users to make phone calls from many locations worldwide. They can be used within range of the mobile telephone site used to house the equipment required to transmit and receive the radio signals from these instruments.

Data Communications

Wireless data communications are an essential component of mobile computing. The various available technologies differ in local availability, coverage range and performance, and in some circumstances, users must be able to employ multiple connection types and switch between them. To simplify the experience for the user, connection manager software can be used, or a mobile VPN deployed to handle the multiple connections as a secure, single virtual network. Supporting technologies include:

> Wi-Fi is a wireless local area network that enables portable computing devices to connect easily to the Internet. Standardized as IEEE 802.11 a,b,g,n, Wi-Fi approaches speeds of some types of wired Ethernet. Wi-Fi has become the de facto standard for access in private homes, within offices, and at public hotspots. Some businesses charge customers a monthly fee for service, while others have begun offering it for free in an effort to increase the sales of their goods.
>
> Cellular data service offers coverage within a range of 10-15 miles from the nearest cell site. Speeds have increased as technologies have evolved, from earlier technologies such as GSM, CDMA and GPRS, to 3G networks such as W-CDMA, EDGE or CDMA2000.

Mobile Satellite Communications may be used where other wireless connections are unavailable, such as in largely rural areas or remote locations. Satellite communications are especially important for transportation, aviation, maritime and military use.

Wireless Sensor Networks are responsible for sensing noise, interference, and activity in data collection networks. This allows us to detect relevant quantities, monitor and collect data, formulate clear user displays, and to perform decision-making functions

Energy Transfer

Wireless energy transfer is a process whereby electrical energy is transmitted from a power source to an electrical load (Computer Load) that does not have a built-in power source, without the use of interconnecting wires. There are two different fundamental methods for wireless energy transfer. They can be transferred using either far-field methods that involve beaming power/lasers, radio or microwave transmissions or near-field using induction. Both methods utilize electromagnetism and magnetic fields.

Medical Technologies

New wireless technologies, such as mobile body area networks (MBAN), have the capability to monitor blood pressure, heart rate, oxygen level and body temperature. The MBAN works by sending low powered wireless signals to receivers that feed into nursing stations or monitoring sites. This technology helps with the intentional and unintentional risk of infection or disconnection that arise from wired connections.

Computer Interface Devices

Answering the call of customers frustrated with cord clutter, many[who?] manufacturers of computer peripherals turned to wireless technology to satisfy their consumer base. Originally these units used bulky, highly local transceivers to mediate between a computer and a keyboard and mouse; however, more recent generations have used small, high-quality devices, some even incorporating Bluetooth. These systems have become so ubiquitous that some users have begun complaining about a lack of wired peripherals.Wireless devices tend to have a slightly slower response time than their wired counterparts; however, the gap is decreasing.

A battery powers computer interface devices such as a keyboard or mouse and send signals to a receiver through a USB port by the way of a radio frequency (RF) receiver. The RF design makes it possible for signals to be transmitted wirelessly and expands the range of efficient use, usually up to 10 feet. Distance, physical obstacles, competing signals, and even human bodies can all degrade the signal quality. Concerns about the security of wireless keyboards arose at the end of 2007, when it was revealed that Microsoft's implementation of encryption in some of its 27 MHz models was highly insecure.

References

- Franconi, N. G.; Bunger, A; Sejdic, E; Mickle, M. H. (2014). "Wireless Communication in Oil and Gas Wells". Energy Technology. 2: 996–1005. doi:10.1002/ente.201402067.

- F.L. Lewis. "Wireless Sensor Networks." Smart Environments: Technologies, Protocols, and Applications, ed. D.J. Cook and S.K. Das, John Wiley, New York, 2004. Automation and robotics research institute. 26 Oct. 2013
- Jones, George. "Future Proof. How Wireless Energy Transfer Will Kill the Power Cable." MaximumPC. 14 Sept. 2010. Web. 26 Oct. 2013.
- Linebaugh, Kate. "Medical Devices in Hospitals go wireless." Online.wsj. The Wall Street Journal. 23 May 2010. Web. 27 Oct. 2013.
- Vilorio, Dennis. "You're a what? Tower Climber" (PDF). Occupational Outlook Quarterly. Archived (PDF) from the original on February 3, 2013. Retrieved December 6, 2013.
- Moser, Max; Schrödel, Philipp (2007-12-05). "27Mhz Wireless Keyboard Analysis Report aka "We know what you typed last summer"" (PDF). Retrieved 6 February 2012.
- "High Speed Internet on the Road". Archived from the original on September 3, 2011. Retrieved September 6, 2011.

2

Networks Related to Wireless Communication

The networks mentioned in this chapter are wireless network, wireless mesh network, personal area network, space network, comparison of wireless data standard and multipath propagation. Wireless networks use wireless connections; they are mostly used in radio communications. The most common examples of wireless networks include cell phone networks, wireless sensor networks and satellite communication networks. This text elucidates the networks related to wireless communication.

Wireless Network

Wireless icon

A wireless network is any type of computer network that uses wireless data connections for connecting network nodes

Wireless networking is a method by which homes, telecommunications networks and enterprise (business) installations avoid the costly process of introducing cables into a building, or as a connection between various equipment locations. Wireless telecommunications networks are generally implemented and administered using radio communication. This implementation takes place at the physical level (layer) of the OSI model network structure.

Examples of wireless networks include cell phone networks, Wireless local networks, wireless sensor networks, satellite communication networks, and terrestrial microwave networks.

Wireless Links

- *Terrestrial microwave* – Terrestrial microwave communication uses Earth-based transmitters and receivers resembling satellite dishes. Terrestrial microwaves are in the low gigahertz range, which limits all communications to line-of-sight. Relay stations are spaced approximately 48 km (30 mi) apart.
- *Communications satellites* – Satellites communicate via microwave radio waves, which are not deflected by the Earth's atmosphere. The satellites are stationed in space, typically in geosynchronous orbit 35,400 km (22,000 mi) above the equator. These Earth-orbiting systems are capable of receiving and relaying voice, data, and TV signals.
- *Cellular and PCS systems* use several radio communications technologies. The systems divide the region covered into multiple geographic areas. Each area has a low-power transmitter or radio relay antenna device to relay calls from one area to the next area.
- *Radio and spread spectrum technologies* – Wireless local area networks use a high-frequency radio technology similar to digital cellular and a low-frequency radio technology. Wireless LANs use spread spectrum technology to enable communication between multiple devices in a limited area. IEEE 802.11 defines a common flavor of open-standards wireless radio-wave technology known as Wifi.
- *Free-space optical communication* uses visible or invisible light for communications. In most cases, line-of-sight propagation is used, which limits the physical positioning of communicating devices.

Types of Wireless Networks

Wireless PAN

Wireless personal area networks (WPANs) interconnect devices within a relatively small area, that is generally within a person's reach. For example, both Bluetooth radio and invisible infrared light provides a WPAN for interconnecting a headset to a laptop. ZigBee also supports WPAN applications. Wi-Fi PANs are becoming commonplace (2010) as equipment designers start to integrate Wi-Fi into a variety of consumer electronic devices. Intel "My WiFi" and Windows 7 "virtual Wi-Fi" capabilities have made Wi-Fi PANs simpler and easier to set up and configure.

Wireless LAN

A wireless local area network (WLAN) links two or more devices over a short distance using a wireless distribution method, usually providing a connection through an access point for internet access. The use of spread-spectrum or OFDM technologies may allow users to move around within a local coverage area, and still remain connected to the network.

Products using the IEEE 802.11 WLAN standards are marketed under the Wi-Fi brand name. Fixed wireless technology implements point-to-point links between computers or networks at two distant locations, often using dedicated microwave or modulated laser light beams over line of sight paths. It is often used in cities to connect networks in two or more buildings without installing a wired link.

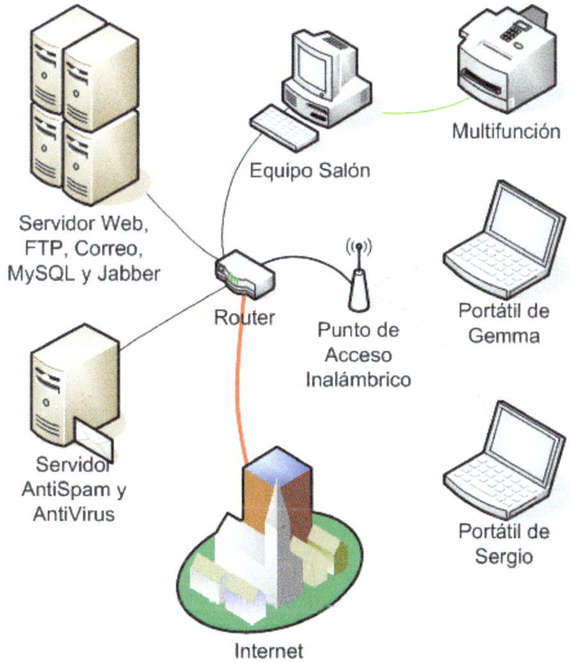

Wireless LANs are often used for connecting to local resources and to the Internet

Wireless Mesh Network

A wireless mesh network is a wireless network made up of radio nodes organized in a mesh topology. Each node forwards messages on behalf of the other nodes. Mesh networks can "self-heal", automatically re-routing around a node that has lost power.

Wireless MAN

Wireless metropolitan area networks are a type of wireless network that connects several wireless LANs.

- WiMAX is a type of Wireless MAN and is described by the IEEE 802.16 standard.

Wireless WAN

Wireless wide area networks are wireless networks that typically cover large areas, such as between neighbouring towns and cities, or city and suburb. These networks can be used to connect branch offices of business or as a public Internet access system. The wireless connections between access points are usually point to point microwave links using parabolic dishes on the 2.4 GHz band, rather than omnidirectional antennas used with smaller networks. A typical system contains base station gateways, access points and wireless bridging relays. Other configurations are mesh systems where each access point acts as a relay also. When combined with renewable energy systems such as photovoltaic solar panels or wind systems they can be stand alone systems.

Global Area Network

A global area network (GAN) is a network used for supporting mobile across an arbitrary number of wireless LANs, satellite coverage areas, etc. The key challenge in mobile communications is handing off user communications from one local coverage area to the next. In IEEE Project 802, this involves a succession of terrestrial wireless LANs.

Space Network

Space networks are networks used for communication between spacecraft, usually in the vicinity of the Earth. The example of this is NASA's Space Network.,

Different Uses

Some examples of usage include cellular phones which are part of everyday wireless networks, allowing easy personal communications. Another example, Intercontinental network systems, use radio satellites to communicate across the world. Emergency services such as the police utilize wireless networks to communicate effectively as well. Individuals and businesses use wireless networks to send and share data rapidly, whether it be in a small office building or across the world.

Properties

General

In a general sense, wireless networks offer a vast variety of uses by both business and home users.

"Now, the industry accepts a handful of different wireless technologies. Each wireless technology is defined by a standard that describes unique functions at both the Physical and the Data Link layers of the OSI model. These standards differ in their specified signaling methods, geographic ranges, and frequency usages, among other things. Such differences can make certain technologies better suited to home networks and others better suited to network larger organizations."

Performance

Each standard varies in geographical range, thus making one standard more ideal than the next depending on what it is one is trying to accomplish with a wireless network. The performance of wireless networks satisfies a variety of applications such as voice and video. The use of this technology also gives room for expansions, such as from 2G to 3G and, most recently, 4G technology, which stands for the fourth generation of cell phone mobile communications standards. As wireless networking has become commonplace, sophistication increases through configuration of network hardware and software, and greater capacity to send and receive larger amounts of data, faster, is achieved.

Space

Space is another characteristic of wireless networking. Wireless networks offer many advantages when it comes to difficult-to-wire areas trying to communicate such as across a street or river, a warehouse on the other side of the premises or buildings that are physically separated but operate as one. Wireless networks allow for users to designate a certain space which the network will be able to communicate with other devices through that network. Space is also created in homes as a result of eliminating clutters of wiring. This technology allows for an alternative to installing physical network mediums such as TPs, coaxes, or fiber-optics, which can also be expensive.

Home

For homeowners, wireless technology is an effective option compared to Ethernet for sharing printers, scanners, and high-speed Internet connections. WLANs help save the cost of installation of cable mediums, save time from physical installation, and also creates mobility for devices connected to the network. Wireless networks are simple and require as few as one single wireless access point connected directly to the Internet via a router.

Wireless Network Elements

The telecommunications network at the physical layer also consists of many interconnected wireline network elements (NEs). These NEs can be stand-alone systems or products that are either supplied by a single manufacturer or are assembled by the service provider (user) or system integrator with parts from several different manufacturers.

Wireless NEs are the products and devices used by a wireless carrier to provide support for the backhaul network as well as a mobile switching center (MSC).

Reliable wireless service depends on the network elements at the physical layer to be protected against all operational environments and applications.

What are especially important are the NEs that are located on the cell tower to the base station (BS) cabinet. The attachment hardware and the positioning of the antenna and associated closures and cables are required to have adequate strength, robustness, corrosion resistance, and resistance against wind, storms, icing, and other weather conditions. Requirements for individual components, such as hardware, cables, connectors, and closures, shall take into consideration the structure to which they are attached.

Difficulties

Interferences

Compared to wired systems, wireless networks are frequently subject to electromagnetic interference. This can be caused by other networks or other types of equipment that generate radio waves that are within, or close, to the radio bands used for communication. Interference can degrade the signal or cause the system to fail.

Absorption and Reflection

Some materials cause absorption of electromagnetic waves, preventing it from reaching the receiver, in other cases, particularly with metallic or conductive materials reflection occurs. This can cause dead zones where no reception is available. Aluminium foiled thermal isolation in modern homes can easily reduce indoor mobile signals by 10 dB frequently leading to complaints about the bad reception of long-distance rural cell signals.

Multipath Fading

In multipath fading two or more different routes taken by the signal, due to reflections, can cause the signal to cancel out at certain locations, and to be stronger in other places (upfade).

Hidden Node Problem

The hidden node problem occurs in some types of network when a node is visible from a wireless access point (AP), but not from other nodes communicating with that AP. This leads to difficulties in media access control.

Shared Resource Problem

The wireless spectrum is a limited resource and shared by all nodes in the range of its transmitters. Bandwidth allocation becomes complex with multiple participating users. Often users are not aware that advertised numbers (e.g., for IEEE 802.11 equipment or LTE networks) are not their capacity, but shared with all other users and thus the individual user rate is far lower. With increasing demand, the capacity crunch is more and more likely to happen. User-in-the-loop (UIL) may be an alternative solution to ever upgrading to newer technologies for over-provisioning.

Capacity

Channel

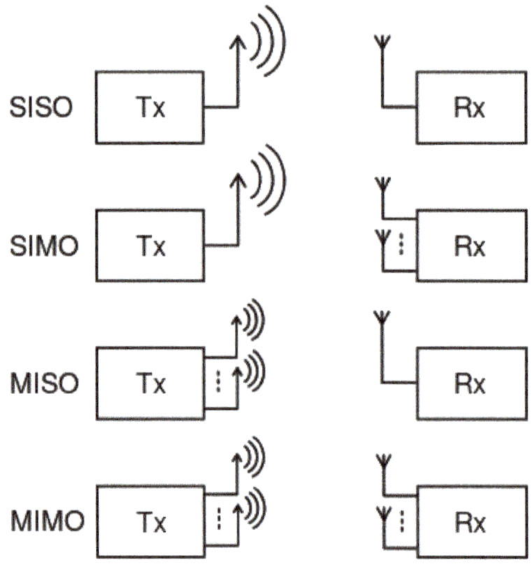

Understanding of SISO, SIMO, MISO and MIMO. Using multiple antennas and transmitting in different frequency channels can reduce fading, and can greatly increase the system capacity.

Shannon's theorem can describe the maximum data rate of any single wireless link, which relates to the bandwidth in hertz and to the noise on the channel.

One can greatly increase channel capacity by using MIMO techniques, where multiple aerials or multiple frequencies can exploit multiple paths to the receiver to achieve much higher throughput – by a factor of the product of the frequency and aerial diversity at each end.

Under Linux, the Central Regulatory Domain Agent (CRDA) controls the setting of channels.

Network

The total network bandwidth depends on how dispersive the medium is (more dispersive medium generally has better total bandwidth because it minimises interference), how many frequencies are available, how noisy those frequencies are, how many aerials are used and whether a directional antenna is in use, whether nodes employ power control and so on.

Cellular wireless networks generally have good capacity, due to their use of directional aerials, and their ability to reuse radio channels in non-adjacent cells. Additionally, cells can be made very small using low power transmitters this is used in cities to give network capacity that scales linearly with population density.

Safety

Wireless access points are also often close to humans, but the drop off in power over distance is fast, following the inverse-square law. The position of the United Kingdom's Health Protection Agency (HPA) is that "...radio frequency (RF) exposures from WiFi are likely to be lower than those from mobile phones." It also saw "...no reason why schools and others should not use WiFi equipment." In October 2007, the HPA launched a new "systematic" study into the effects of WiFi networks on behalf of the UK government, in order to calm fears that had appeared in the media in a recent period up to that time". Dr Michael Clark, of the HPA, says published research on mobile phones and masts does not add up to an indictment of WiFi.

Wireless Mesh Network

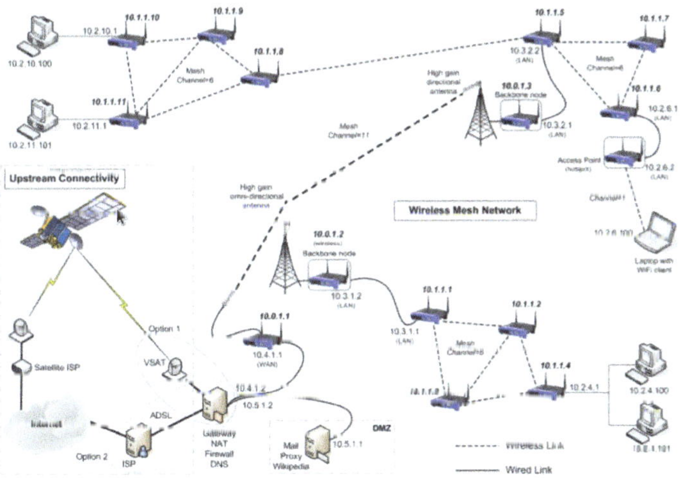

Diagram showing a possible configuration for a wireless mesh network, connected upstream via a VSAT link.

A wireless mesh network (WMN) is a communications network made up of radio nodes organized in a mesh topology. It is also a form of wireless ad hoc network. Wireless mesh networks often consist of mesh clients, mesh routers and gateways. The mesh clients are often laptops, cell phones and other wireless devices while the mesh routers forward traffic to and from the gateways which may, but need not, be connected to the Internet. The coverage area of the radio nodes working as a single network is sometimes called a mesh cloud. Access to this mesh cloud is dependent on the radio nodes working in harmony with each other to create a radio network. A mesh network is reliable and offers redundancy. When one node can no longer operate, the rest of the nodes can still communicate with each other, directly or through one or more intermediate nodes. Wireless mesh networks can self form and self heal. Wireless mesh networks can be implemented with various wireless technologies including 802.11, 802.15, 802.16, cellular technologies and need not be restricted to any one technology or protocol.

History

Architecture

Wireless mesh architecture is a first step towards providing cost effective and dynamic high-bandwidth networks over a specific coverage area. Wireless mesh infrastructure is, in effect, a network of routers minus the cabling between nodes. It's built of peer radio devices that don't have to be cabled to a wired port like traditional WLAN access points (AP) do. Mesh infrastructure carries data over large distances by splitting the distance into a series of short hops. Intermediate nodes not only boost the signal, but cooperatively pass data from point A to point B by making forwarding decisions based on their knowledge of the network, i.e. perform routing. Such an architecture may, with careful design, provide high bandwidth, spectral efficiency, and economic advantage over the coverage area.

Wireless mesh networks have a relatively stable topology except for the occasional failure of nodes or addition of new nodes. The path of traffic, being aggregated from a large number of end users, changes infrequently. Practically all the traffic in an infrastructure mesh network is either forwarded to or from a gateway, while in ad hoc networks or client mesh networks the traffic flows between arbitrary pairs of nodes.

Management

This type of infrastructure can be decentralized (with no central server) or centrally managed (with a central server). Both are relatively inexpensive, and can be very reliable and resilient, as each node needs only transmit as far as the next node. Nodes act as routers to transmit data from nearby nodes to peers that are too far away to reach in a single hop, resulting in a network that can span larger distances. The topology of a mesh network is also reliable, as each node is connected to several other nodes. If one node drops out of the network, due to hardware failure or any other reason, its neighbors can quickly find another route using a routing protocol.

Applications

Mesh networks may involve either fixed or mobile devices. The solutions are as diverse as communication needs, for example in difficult environments such as emergency situations, tunnels, oil rigs, battlefield surveillance, high-speed mobile-video applications on board public transport or real-time racing-car telemetry. An important possible application for wireless mesh networks is VoIP. By using a Quality of Service scheme, the wireless mesh may support local telephone calls to be routed through the mesh.

Some current applications:

- U.S. military forces are now using wireless mesh networking to connect their computers, mainly ruggedized laptops, in field operations.
- Electric meters now being deployed on residences transfer their readings from one to another and eventually to the central office for billing without the need for human meter readers or the need to connect the meters with cables.
- The laptops in the One Laptop per Child program use wireless mesh networking to enable students to exchange files and get on the Internet even though they lack wired or cell phone or other physical connections in their area.
- The 66-satellite Iridium constellation operates as a mesh network, with wireless links between adjacent satellites. Calls between two satellite phones are routed through the mesh, from one satellite to another across the constellation, without having to go through an earth station. This makes for a smaller travel distance for the signal, reducing latency, and also allows for the constellation to operate with far fewer earth stations than would be required for 66 traditional communications satellites.

Operation

The principle is similar to the way packets travel around the wired Internet – data will hop from one device to another until it eventually reaches its destination. Dynamic routing algorithms implemented in each device allow this to happen. To implement such dynamic routing protocols, each device needs to communicate routing information to other devices in the network. Each device then determines what to do with the data it receives – either pass it on to the next device or keep it, depending on the protocol. The routing algorithm used should attempt to always ensure that the data takes the most appropriate (fastest) route to its destination.

Multi-radio Mesh

Multi-radio mesh refers to a unique pair of dedicated radios on each end of the link. This means there is a unique frequency used for each wireless hop and thus a dedicated CSMA collision domain. This is a true mesh link where you can achieve maximum performance without bandwidth degradation in the mesh and without adding latency. Thus voice and video applications work just as they would on a wired Ethernet network. In true 802.11 networks, there is no concept of a mesh. There are only APs and Stations. A multi-radio wireless mesh node will dedicate one of the radios to act as a station, and connect to a neighbor node AP radio.

Research Topics

One of the more often cited papers on Wireless Mesh Networks identified the following areas as open research problems in 2005

- New modulation scheme
    - In order to achieve higher transmission rate, new wideband transmission schemes other than OFDM and UWB are needed.

- Advanced antenna processing
  - Advanced antenna processing including directional, smart and multiple antenna technologies is further investigated, since their complexity and cost are still too high for wide commercialization.
- Flexible spectrum management
  - Tremendous efforts on research of frequency-agile techniques are being performed for increased efficiency.
- Cross-layer optimization
  - Cross-layer research is a popular current research topic where information is shared between different communications layers in order to increase the knowledge and current state of the network. This could enable new and more efficient protocols to be developed. A joint protocol which combines various design problems like routing, scheduling, channel assignment etc. can achieve higher performance since it is proven that these problems are strongly co-related. It is important to note that careless cross-layer design could lead to code which is difficult to maintain and extend.
- Software-defined wireless networking
  - Centralized, distributed, or hybrid? - In a new SDN architecture for WDNs is explored that eliminates the need for multi-hop flooding of route information and therefore enables WDNs to easily expand. The key idea is to split network control and data forwarding by using two separate frequency bands. The forwarding nodes and the SDN controller exchange link-state information and other network control signaling in one of the bands, while actual data forwarding takes place in the other band.
- Security
  - A WMN can be seen as a group of nodes (clients or routers) that cooperate to provide connectivity. Such an open architecture, where clients serve as routers to forward data packets, is exposed to many types of attacks that can interrupt the whole network and cause denial of service (DoS) or Distributed Denial of Service (DDoS).

Personal Area Network

A personal area network (PAN) is a computer network used for data transmission amongst devices such as computers, telephones, tablets and personal digital assistants. PANs can be used for communication amongst the personal devices themselves (interpersonal communication), or for connecting to a higher level network and the Internet (an uplink) where one "master" device takes up the role as internet router.

A wireless personal area network (WPAN) is a low-powered PAN carried over a short-distance wireless network technology such as:

- INSTEON
- IrDA
- Wireless USB
- Bluetooth
- Z-Wave
- ZigBee
- Body Area Network

The reach of a WPAN varies from a few centimeters to a few meters. A PAN may also be carried over wired computer buses such as USB and FireWire.

Although a (secured) Wi-Fi tethering connection could be used by only one single user it is not considered to be a PAN.

Wired PAN Connection

The data cable is an example of the above PAN. This is also a Personal Area Network because that connection is for the users personal use. PAN is used for personal use only.

Wireless Personal Area Network

A wireless personal area network (WPAN) is a personal area network—a network for interconnecting devices centered on an individual person's workspace—in which the connections are wireless. Wireless PAN is based on the standard IEEE 802.15. The two kinds of wireless technologies used for WPAN are Bluetooth and Infrared Data Association.

A WPAN could serve to interconnect all the ordinary computing and communicating devices that many people have on their desk or carry with them today; or it could serve a more specialized purpose such as allowing the surgeon and other team members to communicate during an operation.

A key concept in WPAN technology is known as "plugging in". In the ideal scenario, when any two WPAN-equipped devices come into close proximity (within several meters of each other) or within a few kilometers of a central server, they can communicate as if connected by a cable. Another important feature is the ability of each device to lock out other devices selectively, preventing needless interference or unauthorized access to information.

The technology for WPANs is in its infancy and is undergoing rapid development. Proposed operating frequencies are around 2.4 GHz in digital modes. The objective is to facilitate seamless operation among home or business devices and systems. Every device in a WPAN will be able to plug into any other device in the same WPAN, provided they are within physical range of one another. In addition, WPANs worldwide will be interconnected. Thus, for example, an archeologist on site in Greece might use a PDA to directly access databases at the University of Minnesota in Minneapolis, and to transmit findings to that database.

Bluetooth

Bluetooth uses short-range radio waves over distances up to approximately 10 metres. For example, Bluetooth devices such as a keyboards, pointing devices, audio head sets, printers may connect to personal digital assistants (PDAs), cell phones, or computers wirelessly.

A Bluetooth PAN is also called a *piconet* (combination of the prefix "pico," meaning very small or one trillionth, and network), and is composed of up to 8 active devices in a master-slave relationship (a very large number of devices can be connected in "parked" mode). The first Bluetooth device in the piconet is the master, and all other devices are slaves that communicate with the master. A piconet typically has a range of 10 metres (33 ft), although ranges of up to 100 metres (330 ft) can be reached under ideal circumstances.

Infrared Data Association

Infrared Data Association (IrDA) uses infrared light, which has a frequency below the human eye's sensitivity. Infrared in general is used, for instance, in TV remotes. Typical WPAN devices that use IrDA include printers, keyboards, and other serial data interfaces.

Space Network

Space Network (SN) is a NASA program that combines space and ground elements to support spacecraft communications in Earth vicinity. The SN Project Office at Goddard Space Flight Center (GSFC) manages the SN, which consists of:

- The geosynchronous Tracking and Data Relay Satellites (TDRS),
- Supporting ground terminal systems,
- The Bilateration Ranging and Transponder System (BRTS),
- Merritt Island Launch Annex (MILA) relay,
- Network Control Center Data System (NCCDS).

Satellite Generations

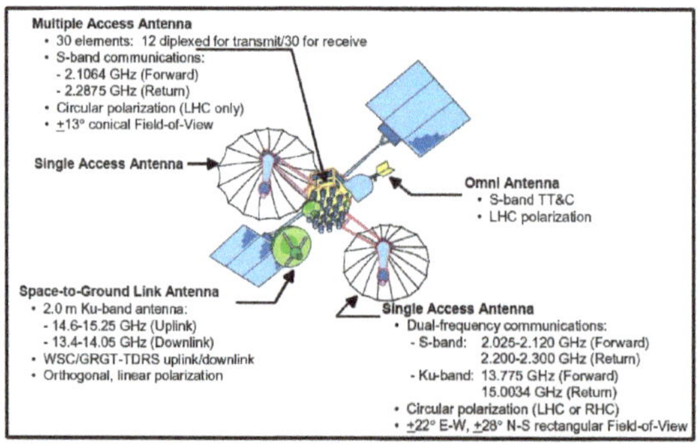

First Generation Tracking and Data Relay Satellite (F1-F7)

First Generation Tracking and Data Relay Satellite (F1-F7)

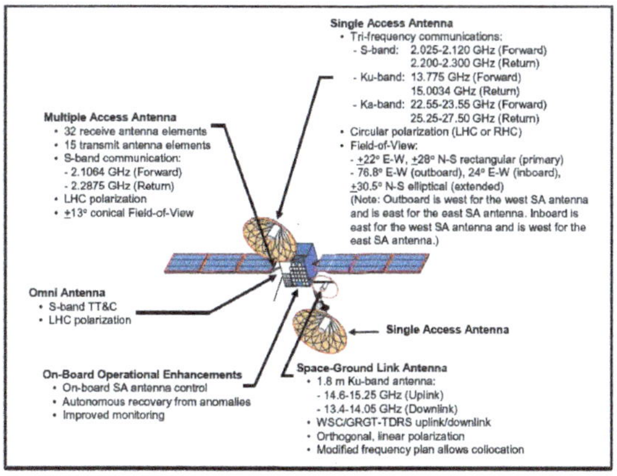

Second Generation Tracking and Data Relay Satellite F8-F10

Second Generation Tracking and Data Relay Satellite (F8-F10 also known as H, I, J)

Tracking and Data Relay Satellite (TDRS) currently consists of first generation (F1-F7), and second generation (F8-F10) satellites.

The space segment of the SN consists of up to six operational relay satellites in geosynchronous orbit. These communications satellites are allocated longitudes for relaying forward and return service signals to and from customers, any entity with an Earth-orbiting satellite that has an agreement with SN to use its communications services, for data transfer and tracking. An additional TDRS, F1, provides dedicated support to the National Science Foundation (NSF) through the use of the WSC Alternate Relay Terminal (WART). Additional spare TDRSs may be in geosynchronous orbit.

All first generation TDRSs (F1-F7, also known as TDRS A-G) carry functionally identical payloads and all second generation TDRSs (F8-F10, also known as TDRS H-J) carry functionally identical payloads.

A third generation, TDRS K and L, are planned for launch 2012-2013. See TDRS launch history.

Click on the figures to the right, which identify the pertinent communications components and associated parameters of the orbiting relay platforms.

Coverage

For spacecraft operating in a low earth orbit (LEO) 73 km to 3000 km altitude, the SN is capable of providing tracking and data acquisition services over 100% of the spacecraft's orbit. Spacecraft sent to more distant or exotic destinations rely on either Deep Space Network or their own custom, dedicated networks.

Comparison of Wireless Data Standards

A wide variety of different wireless data technologies exist, some in direct competition with one another, others designed for specific applications. Wireless technologies can be evaluated by a variety of different metrics of which some are described in this entry.

Standards can be grouped as follows in increasing range order:

Personal Area Network (PAN) systems are intended for short range communication between devices typically controlled by a single person. Some examples include wireless headsets for mobile phones or wireless heart rate sensors communicating with a wrist watch. Some of these technologies include standards such as ANT UWB, Bluetooth, ZigBee, and Wireless USB.

Wireless Sensor Networks (WSN / WSAN) are, generically, networks of low-power, low-cost devices that interconnect wirelessly to collect, exchange, and sometimes act-on data collected from their physical environments - "sensor networks". Nodes typically connect in a star or mesh topology. While most individual nodes in a WSAN are expected to have limited range (Bluetooth, ZigBee, 6LoWPAN, etc.), particular nodes may be capable of more expansive communications (Wi-Fi, Cellular networks, etc.) and any individual WSAN can span a wide geographical range. An example of a WSAN would be a collection of sensors arranged throughout an agricultural facility to monitor soil moisture levels, report the data back to a computer in the main office for analysis and trend modeling, and maybe turn on automatic watering spigots if the level is too low.

For wider area communications, Wireless Local Area Network (WLAN) is used. WLANs are often known by their commercial product name Wi-Fi. These systems are used to provide wireless access to other systems on the local network such as other computers, shared printers, and other such devices or even the internet. Typically a WLAN offers much better speeds and delays within the local network than an average consumer's Internet access. Older systems that provide WLAN functionality include DECT and HIPERLAN. These however are no longer in widespread use. One typical characteristic of WLANs is that they are mostly very local, without the capability of seamless movement from one network to another.

Cellular networks or WAN are designed for city-wide/national/global coverage areas and seamless mobility from one access point (often defined as a Base Station) to another allowing seamless coverage for very wide areas. Cellular network technologies are often split into 2nd generation 2G, 3G and 4G networks. Originally 2G networks were voice centric or even voice only digital cellular systems (as opposed to the analog 1G networks). Typical 2G standards include GSM and IS-95 with extensions via GPRS, EDGE and 1xRTT, providing Internet access to users of originally voice centric 2G networks. Both EDGE and 1xRTT are 3G standards, as defined by the ITU, but are usually marketed as 2.9G due to their comparatively low speeds and high delays when compared to true 3G technologies.

True 3G systems such as EV-DO, W-CDMA (including HSPA) provide combined circuit switched and packet switched data and voice services from the outset, usually at far better data rates than 2G networks with their extensions. All of these services can be used to provide combined mobile voice access and Internet access at remote locations.

4G networks provide even higher bitrates and many architectural improvements, which are not necessarily visible to the consumer. The current 4G systems that are deployed widely are HSPA+, WIMAX and LTE. The latter two are pure packet based networks without traditional voice circuit capabilities. These networks provide voice services via VoIP.

Some systems are designed for point-to-point line-of-sight communications, once two such nodes get too far apart they can no longer communicate. Other systems are designed to form a wireless mesh network using one of a variety of routing protocols. In a mesh network, when nodes get too far apart to communicate directly, they can still communicate indirectly through intermediate nodes.

Standards

The following standards are included in this comparison.

Wireless Wide Area Network (WWAN)

- EDGE
- EV-DO x1 Rev 0, Rev A, Rev B and x3 standards.
- Flash-OFDM: FLASH(Fast Low-latency Access with Seamless Handoff)-OFDM (Orthogonal Frequency Division Multiplexing)
- GPRS
- HSPA D and U standards.
- Lorawan
- LTE
- RTT
- UMTS over W-CDMA
- UMTS-TDD
- WiMAX: 802.16 standard

Wireless Local Area Network (WLAN)

- Wi-Fi: 802.11a, 802.11b, 802.11g, 802.11n, 802.11ac standards.

Wireless Personal Area Network (WPAN) and Most Wireless Sensor Actor Networks (WSAN)

- 6LoWPAN
- Bluetooth V4.0 with standard protocol and with low energy protocol
- IEEE 802.15.4-2006 (low-level protocol definitions corresponding to the OSI model physical and link layers. ZigBee, 6LoWPAN, etc. build upward in the protocol stack and correspond to the network and transport layers.)
- Thread (network protocol)
- UWB
- Wireless USB
- ZigBee

Overview

Comparison of mobile Internet access methods						
Common Name	**Family**	**Primary Use**	**Radio Tech**	**Downstream (Mbit/s)**	**Upstream (Mbit/s)**	**Notes**
HSPA+	3GPP	3G Data	CDMA/FDD MIMO	21 42 84 672	5.8 11.5 22 168	HSPA+ is widely deployed. Revision 11 of the 3GPP states that HSPA+ is expected to have a throughput capacity of 672 Mbit/s.
LTE	3GPP	General 4G	OFDMA/MIMO/ SC-FDMA	100 Cat3 150 Cat4 300 Cat5 (in 20 MHz FDD)	50 Cat3/4 75 Cat5 (in 20 MHz FDD)	LTE-Advanced update expected to offer peak rates up to 1 Gbit/s fixed speeds and 100 Mb/s to mobile users.
WiMax rel 1	802.16	WirelessMAN	MIMO-SOFDMA	37 (10 MHz TDD)	17 (10 MHz TDD)	With 2x2 MIMO.
WiMax rel 1.5	802.16-2009	WirelessMAN	MIMO-SOFDMA	83 (20 MHz TDD) 141 (2x20 MHz FDD)	46 (20 MHz TDD) 138 (2x20 MHz FDD)	With 2x2 MIMO.Enhanced with 20 MHz channels in 802.16-2009
WiMAX rel 2	802.16m	WirelessMAN	MIMO-SOFDMA	2x2 MIMO 110 (20 MHz TDD) 183 (2x20 MHz FDD) 4x4 MIMO 219 (20 MHz TDD) 365 (2x20 MHz FDD)	2x2 MIMO 70 (20 MHz TDD) 188 (2x20 MHz FDD) 4x4 MIMO 140 (20 MHz TDD) 376 (2x20 MHz FDD)	Also, low mobility users can aggregate multiple channels to get a download throughput of up to 1 Gbit/s
Flash-OFDM	Flash-OFDM	Mobile Internet mobility up to 200 mph (350 km/h)	Flash-OFDM	5.3 10.6 15.9	1.8 3.6 5.4	Mobile range 30 km (18 miles) extended range 55 km (34 miles)
HIPERMAN	HIPERMAN	Mobile Internet	OFDM	56.9		

Wi-Fi	802.11 (11n)	Mobile Internet	OFDM/MIMO	288.8 (using 4x4 configuration in 20 MHz bandwidth) or 600 (using 4x4 configuration in 40 MHz bandwidth)		Antenna, RF front end enhancements and minor protocol timer tweaks have helped deploy long range P2P networks compromising on radial coverage, throughput and/or spectra efficiency (310 km & 382 km)
iBurst	802.20	Mobile Internet	HC-SDMA/TDD/MIMO	95	36	Cell Radius: 3–12 km Speed: 250 km/h Spectral Efficiency: 13 bits/s/Hz/cell Spectrum Reuse Factor: "1"
EDGE Evolution	GSM	Mobile Internet	TDMA/FDD	1.6	0.5	3GPP Release 7
UMTS W-CDMA HSPA (HSDPA+HSUPA)	UMTS/3GSM	General 3G	CDMA/FDD CDMA/FDD/MIMO	0.384 14.4	0.384 5.76	HSDPA is widely deployed. Typical downlink rates today 2 Mbit/s, ~200 kbit/s uplink; HSPA+ downlink up to 56 Mbit/s.
UMTS-TDD	UMTS/3GSM	Mobile Internet	CDMA/TDD	16		Reported speeds according to IPWireless using 16QAM modulation similar to HSDPA+HSUPA
EV-DO Rel. 0 EV-DO Rev.A EV-DO Rev.B	CDMA2000	Mobile Internet	CDMA/FDD	2.45 3.1 4.9xN	0.15 1.8 1.8xN	Rev B note: N is the number of 1.25 MHz carriers used. EV-DO is not designed for voice, and requires a fallback to 1xRTT when a voice call is placed or received.

Notes: All speeds are theoretical maximums and will vary by a number of factors, including the use of external antennas, distance from the tower and the ground speed (e.g. communications on a train may be poorer than when standing still). Usually the bandwidth is shared between several terminals. The performance of each technology is determined by a number of constraints, including the spectral efficiency of the technology, the cell sizes used, and the amount of spectrum available.

For more comparison tables, see bit rate progress trends, comparison of mobile phone standards, spectral efficiency comparison table and OFDM system comparison table.

Peak Bit Rate and Throughput

When discussing throughput, there is often a distinction between the peak data rate of the physical layer, the theoretical maximum data throughput and typical throughput.

The peak bit rate of the standard is the net bit rate provided by the physical layer in the fastest transmission mode (using the fastest modulation scheme and error code), excluding forward error correction coding and other physical layer overhead.

The theoretical maximum throughput for end user is clearly lower than the peak data rate due to higher layer overheads. Even this is never possible to achieve unless the test is done under perfect laboratory conditions.

The typical throughput is what users have experienced most of the time when well within the usable range to the base station. The typical throughput is hard to measure, and depends on many protocol issues such as transmission schemes (slower schemes are used at longer distance from the access point due to better redundancy), packet retransmissions and packet size. The typical throughput is often even lower because of other traffic sharing the same network or cell, interference or even the fixed line capacity from the base station onwards being limited.

Note that these figures cannot be used to predict the performance of any given standard in any given environment, but rather as benchmarks against which actual experience might be compared.

Bit rate (Mbit/s)				
Standard	**Peak Downlink**	**Peak Uplink**	**Approximate Maximum Range in Meters**	**Typical Downlink throughput**
CDMA RTT 1x	0.3072	0.1536	29000	0.125
CDMA EV-DO Rev. 0	2.4580	0.1536	29000	1
CDMA EV-DO Rev. A	3.1	1.8	29000	2
CDMA EV-DO Rev. B	4.9	1.8	29000	
GSM GPRS Class 10	0.0856	0.0428	26000	0.014
GSM EDGE type 2	0.4736	0.4736	26000	0.034

GSM EDGE Evolution	1.8944	0.9472	26000	
UMTS W-CDMA R99	0.3840	0.3840	29000	0.195
UMTS W-CDMA HSDPA	14.4	0.3840	200000	2
UMTS W-CDMA HSUPA	14.4	5.76	200000	
UMTS W-CDMA HSPA+	168	22	200000	
UMTS-TDD	16	16		
LTE	326.4	86.4		
iBurst: iBurst	24	8	12000	>2
Flash-OFDM: Flash-OFDM	5.3	1.8	29000	avg 2.5
WiMAX: 802.16e	70	70	6400	>10
WiFi: 802.11a	54	54	30	20
WiFi: 802.11b	11	11	30	5
WiFi: 802.11g	54	54	30	20
WiFi: 802.11n	600	600	50	
WiFi: 802.11ac	1,300	1,300		

- **Downlink** is the throughput from the base station to the user handset or computer.
- **Uplink** is the throughput from the user handset or computer to the base station.
- **Range** is the maximum range possible to receive data at 25% of the typical rate.

Multipath Propagation

In wireless telecommunications, multipath is the propagation phenomenon that results in radio signals reaching the receiving antenna by two or more paths. Causes of multipath include atmospheric ducting, ionospheric reflection and refraction, and reflection from water bodies and terrestrial objects such as mountains and buildings.

Multipath causes multipath interference including constructive and destructive interference, and phase shifting of the signal. Destructive interference causes fading. Where the magnitudes of the signals arriving by the various paths have a distribution known as the Rayleigh distribution, this is known as Rayleigh fading. Where one component (often, but not necessarily, a line of sight component) dominates, a Rician distribution provides a more accurate model, and this is known as Rician fading.

Examples

In facsimile and (analog) television transmission, multipath causes jitter and ghosting, seen as a faded duplicate image to the right of the main image. Ghosts occur when transmissions bounce off a mountain or other large object, while also arriving at the antenna by a shorter, direct route, with the receiver picking up two signals separated by a delay.

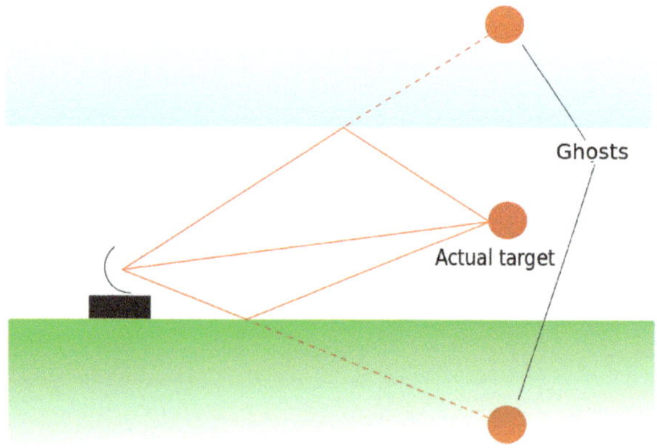

Radar multipath echoes from an actual target cause ghosts to appear.

In radar processing, multipath causes ghost targets to appear, deceiving the radar receiver. These ghosts are particularly bothersome since they move and behave like the normal targets (which they echo), and so the receiver has difficulty in isolating the correct target echo. These problems can be overcome by incorporating a ground map of the radar's surroundings and eliminating all echoes which appear to originate below ground or above a certain height.

In digital radio communications (such as GSM) multipath can cause errors and affect the quality of communications. The errors are due to intersymbol interference (ISI). Equalisers are often used to correct the ISI. Alternatively, techniques such as orthogonal frequency division modulation and rake receivers may be used.

GPS error due to multipath

In a Global Positioning System receiver, Multipath Effect can cause a stationary receiver's output to indicate as if it were randomly jumping about or creeping. When the unit is moving the jumping or creeping may be hidden, but it still degrades the displayed accuracy of location and speed.

In wired media

Multipath propagation may also happen in wired media, especially where impedance mismatch causes signal reflection. A well-known example is power line communication.

High-speed power line communication systems usually employ multi-carrier modulations (such as OFDM or Wavelet OFDM) to avoid the intersymbol interference that multipath propagation would cause.

The ITU-T G.hn standard provides a way to create a high-speed (up to 1 Gigabit/s) local area network using existing home wiring (power lines, phone lines and coaxial cables). G.hn uses OFDM with a cyclic prefix to avoid ISI. Because multipath propagation behaves differently in each kind of wire, G.hn uses different OFDM parameters (OFDM symbol duration, Guard Interval duration) for each media.

Mathematical Modeling

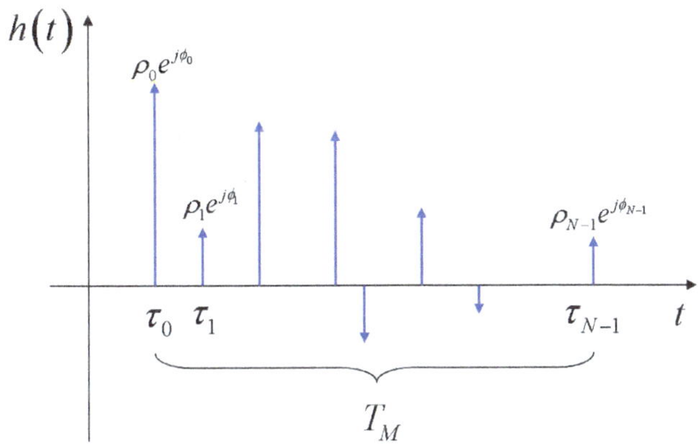

Mathematical model of the multipath impulse response.

The mathematical model of the multipath can be presented using the method of the impulse response used for studying linear systems.

Suppose you want to transmit a single, ideal Dirac pulse of electromagnetic power at time 0, i.e.

$$x(t) = \delta(t)$$

At the receiver, due to the presence of the multiple electromagnetic paths, more than one pulse will be received (we suppose here that the channel has infinite bandwidth, thus the pulse shape is not modified at all), and each one of them will arrive at different times. In fact, since the electromagnetic signals travel at the speed of light, and since every path has a geometrical length possibly different from that of the other ones, there are different air travelling times (consider that, in free space, the light takes 3 μs to cross a 1 km span). Thus, the received signal will be expressed by

$$y(t) = h(t) = \sum_{n=0}^{N-1} \rho_n e^{j\phi_n} \delta(t - \tau_n)$$

where N is the number of received impulses (equivalent to the number of electromagnetic paths, and possibly very large), τ_n is the time delay of the generic n^{th} impulse, and $\rho_n e^{j\phi_n}$ represent the complex amplitude (i.e., magnitude and phase) of the generic received pulse. As a consequence, $y(t)$ also represents the impulse response function $h(t)$ of the equivalent multipath model.

More in general, in presence of time variation of the geometrical reflection conditions, this impulse response is time varying, and as such we have

$$\tau_n = \tau_n(t)$$

$$\rho_n = \rho_n(t)$$

$$\phi_n = \phi_n(t)$$

Very often, just one parameter is used to denote the severity of multipath conditions: it is called the multipath time, T_M, and it is defined as the time delay existing between the first and the last received impulses

$$T_M = \tau_{N-1} - \tau_0$$

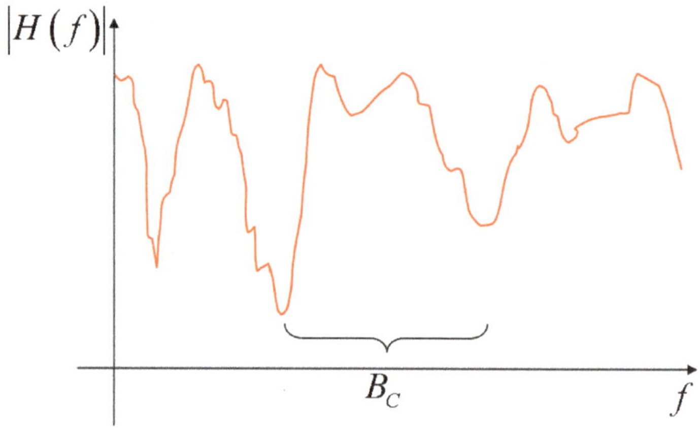

Mathematical model of the multipath channel transfer function.

In practical conditions and measurement, the multipath time is computed by considering as last impulse the first one which allows to receive a determined amount of the total transmitted power (scaled by the atmospheric and propagation losses), e.g. 99%.

Keeping our aim at linear, time invariant systems, we can also characterize the multipath phenomenon by the channel transfer function $H(f)$, which is defined as the continuous time Fourier transform of the impulse response $h(t)$

$$H(f) = \mathfrak{F}(h(t)) = \int_{-\infty}^{+\infty} h(t)e^{-j2\pi ft}dt = \sum_{n=0}^{N-1} \rho_n e^{j\phi_n} e^{-j2\pi f\tau_n}$$

where the last right-hand term of the previous equation is easily obtained by remembering that the Fourier transform of a Dirac pulse is a complex exponential function, an eigenfunction of every linear system.

The obtained channel transfer characteristic has a typical appearance of a sequence of peaks and valleys (also called *notches*); it can be shown that, on average, the distance (in Hz) between two consecutive valleys (or two consecutive peaks), is roughly inversely proportional to the multipath time. The so-called coherence bandwidth is thus defined as

$$B_C \approx \frac{1}{T_M}$$

For example, with a multipath time of 3 μs (corresponding to a 1 km of added on-air travel for the last received impulse), there is a coherence bandwidth of about 330 kHz.

References

- Guowang Miao, Jens Zander, Ki Won Sung, and Ben Slimane, Fundamentals of Mobile Data Networks, Cambridge University Press, ISBN 1107143217, 2016.
- Dean Tamara (2010). Network+ Guide to Networks (5th ed.). Boston: Cengage Learning. ISBN 978-1-4239-0245-4.

3

Radio: An Overview

This section is an overview of the subject matter incorporating all the major aspects of a radio. Some of the topics discussed in this text are radio propagation, radio frequency, skywave and radio broadcasting. The pattern in which radio waves travel is known as radio propagation whereas radio frequency is the frequency that is used for communication and lies in between 3 kHZ to 300 GHz. The topics discussed herein are of great importance to broaden the existing knowledge on radio.

Radio

Radio is the technology of using radio waves to carry information, such as sound, by systematically modulating some property of electromagnetic energy waves transmitted through space, such as their amplitude, frequency, phase, or pulse width. When radio waves strike an electrical conductor, the oscillating fields induce an alternating current in the conductor. The information in the waves can be extracted and transformed back into its original form.

The Alexandra Palace, here: mast of the broadcasting station

Radio systems need a transmitter to modulate (change) some property of the energy produced to impress a signal on it, for example using amplitude modulation or angle modulation (which can be frequency modulation or phase modulation). Radio systems also need an antenna to convert electric currents into radio waves, and vice versa. An antenna can be used for both transmitting and receiving. The electrical resonance of tuned circuits in radios allow individual stations to be selected. The electromagnetic wave is intercepted by a tuned receiving antenna. A radio receiver receives its input from an antenna and converts it into a form usable for the consumer, such as sound, pictures, digital data, measurement values, navigational positions, etc. Radio frequencies occupy the range from a 3 kHz to 300 GHz, although commercially important uses of radio use only a small part of this spectrum.

Classic radio receiver dial

A radio communication system sends signals by radio. The radio equipment involved in communication systems includes a transmitter and a receiver, each having an antenna and appropriate terminal equipment such as a microphone at the transmitter and a loudspeaker at the receiver in the case of a voice-communication system.

Etymology

Following Heinrich Hertz's establishment of the existence of electromagnetic radiation in the late 1880s, a variety of terms were initially used for the phenomenon, with early descriptions of the radiation itself including "Hertzian waves", "electric waves", and "ether waves", while phrases describing its use in communications included "spark telegraphy", "space telegraphy", "aerography" and, eventually and most commonly, "wireless telegraphy". However, "wireless" included a broad variety of related electronic technologies, including electrostatic induction, electromagnetic induction and aquatic and earth conduction, so there was a need for a more precise term referring exclusively to electromagnetic radiation.

The first use of *radio-* in conjunction with electromagnetic radiation appears to have been by French physicist Édouard Branly, who in 1890 developed a version of a coherer receiver he called a *radio-conducteur*. The radio- prefix was later used to form additional descriptive compound and hyphenated words, especially in Europe, for example, in early 1898 the British publication *The Practical Engineer* included a reference to "the radiotelegraph" and "radiotelegraphy", while the

French text of both the 1903 and 1906 Berlin Radiotelegraphic Conventions includes the phrases *radiotélégraphique* and *radiotélégrammes*.

The use of "radio" as a standalone word dates back to at least December 30, 1904, when instructions issued by the British Post Office for transmitting telegrams specified that "The word 'Radio'... is sent in the Service Instructions". This practice was universally adopted, and the word "radio" introduced internationally, by the 1906 Berlin Radiotelegraphic Convention, which included a Service Regulation specifying that "Radiotelegrams shall show in the preamble that the service is 'Radio'".

The switch to "radio" in place of "wireless" took place slowly and unevenly in the English-speaking world. Lee de Forest helped popularize the new word in the United States—in early 1907 he founded the DeForest Radio Telephone Company, and his letter in the June 22, 1907 *Electrical World* about the need for legal restrictions warned that "Radio chaos will certainly be the result until such stringent regulation is enforced". The United States Navy would also play a role. Although its translation of the 1906 Berlin Convention used the terms "wireless telegraph" and "wireless telegram", by 1912 it began to promote the use of "radio" instead. The term started to become preferred by the general public in the 1920s with the introduction of broadcasting. ("Broadcasting" is based upon an agricultural term meaning roughly "scattering seeds widely".) British Commonwealth countries continued to commonly use the term "wireless" until the mid-20th century, though the magazine of the British Broadcasting Corporation in the UK has been called Radio Times since its founding in the early 1920s.

In recent years the more general term "wireless" has gained renewed popularity, even for devices using electromagnetic radiation, through the rapid growth of short-range computer networking, e.g., Wireless Local Area Network (WLAN), Wi-Fi, and Bluetooth, as well as mobile telephony, e.g., GSM and UMTS cell phones. Today, the term "radio" specifies the transceiver device or chip, whereas "wireless" refers to the lack of physical connections; thus equipment employs embedded *radio* transceivers, but operates as *wireless* devices over *wireless* sensor networks.

Processes

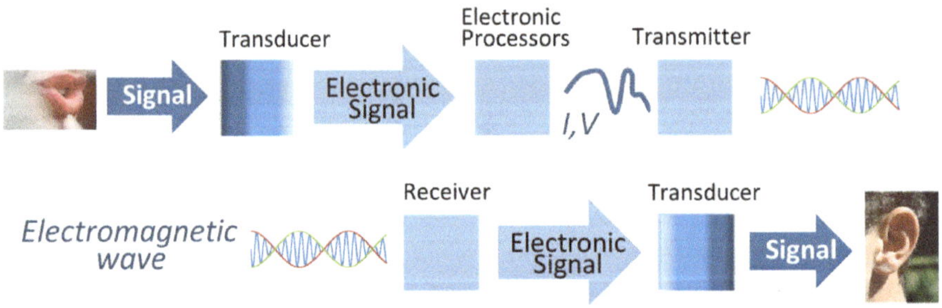

Radio communication. Information such as sound is converted by a transducer such as a microphone to an electrical signal, which modulates a radio wave sent from a transmitter. A receiver intercepts the radio wave and extracts the information-bearing electronic signal, which is converted back using another transducer such as a speaker.

Radio systems used for communication have the following elements. With more than 100 years of development, each process is implemented by a wide range of methods, specialised for different communications purposes.

Transmitter and Modulation

Each system contains a transmitter, This consists of a source of electrical energy, producing alternating current of a desired frequency of oscillation. The transmitter contains a system to modulate (change) some property of the energy produced to impress a signal on it. This modulation might be as simple as turning the energy on and off, or altering more subtle properties such as amplitude, frequency, phase, or combinations of these properties. The transmitter sends the modulated electrical energy to a tuned resonant antenna; this structure converts the rapidly changing alternating current into an electromagnetic wave that can move through free space (sometimes with a particular polarization).

Amplitude modulation of a carrier wave works by varying the strength of the transmitted signal in proportion to the information being sent. For example, changes in the signal strength can be used to reflect the sounds to be reproduced by a speaker, or to specify the light intensity of television pixels. It was the method used for the first audio radio transmissions, and remains in use today. "AM" is often used to refer to the medium wave broadcast band, but it is used in various radiotelephone services such as the Citizen Band, amateur radio and especially in aviation, due to its ability to be received under very weak signal conditions and its immunity to capture effect, allowing more than one signal to be heard simultaneously.

Frequency modulation varies the frequency of the carrier. The instantaneous frequency of the carrier is directly proportional to the instantaneous value of the input signal. FM has the "capture effect" whereby a receiver only receives the strongest signal, even when others are present. Digital data can be sent by shifting the carrier's frequency among a set of discrete values, a technique known as frequency-shift keying. FM is commonly used at Very high frequency (VHF) radio frequencies for high-fidelity broadcasts of music and speech. Analog TV sound is also broadcast using FM.

Angle modulation alters the instantaneous phase of the carrier wave to transmit a signal. It may be either FM or phase modulation (PM).

Antenna

Rooftop television antennas. Yagi-Uda antennas like these six are widely used at VHF and UHF frequencies.

An *antenna* (or *aerial*) is an electrical device which converts electric currents into radio waves, and vice versa. It is usually used with a radio transmitter or radio receiver. In transmission, a radio transmitter supplies an electric current oscillating at radio frequency (i.e. high frequency AC) to the antenna's terminals, and the antenna radiates the energy from the current as electromagnetic waves (radio waves). In reception, an antenna intercepts some of the power of an electromagnetic wave in order to produce a tiny voltage at its terminals, that is applied to a receiver to be amplified. Some antennas can be used for both transmitting and receiving, even simultaneously, depending on the connected equipment.

Propagation

Once generated, electromagnetic waves travel through space either directly, or have their path altered by reflection, refraction or diffraction. The intensity of the waves diminishes due to geometric dispersion (the inverse-square law); some energy may also be absorbed by the intervening medium in some cases. Noise will generally alter the desired signal; this electromagnetic interference comes from natural sources, as well as from artificial sources such as other transmitters and accidental radiators. Noise is also produced at every step due to the inherent properties of the devices used. If the magnitude of the noise is large enough, the desired signal will no longer be discernible; the signal-to-noise ratio is the fundamental limit to the range of radio communications.

Resonance

Electrical resonance of tuned circuits in radios allow individual stations to be selected. A resonant circuit will respond strongly to a particular frequency, and much less so to differing frequencies. This allows the radio receiver to discriminate between multiple signals differing in frequency.

Receiver and Demodulation

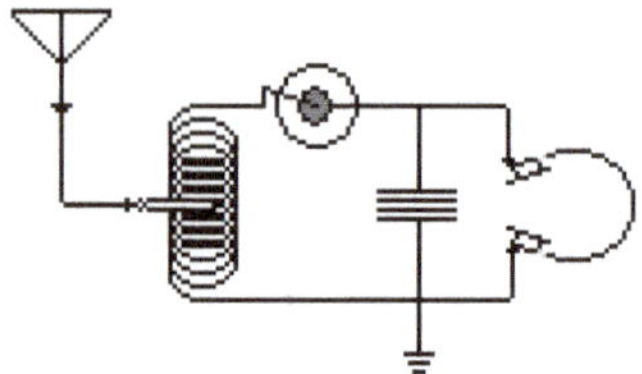

A crystal receiver, consisting of an antenna, adjustable electromagnetic coil, crystal rectifier, capacitor, headphones and ground connection.

The electromagnetic wave is intercepted by a tuned receiving antenna; this structure captures some of the energy of the wave and returns it to the form of oscillating electrical currents. At the receiver, these currents are demodulated, which is conversion to a usable signal form by a detector sub-system. The receiver is "tuned" to respond preferentially to the desired signals, and reject undesired signals.

Early radio systems relied entirely on the energy collected by an antenna to produce signals for the operator. Radio became more useful after the invention of electronic devices such as the vacuum tube and later the transistor, which made it possible to amplify weak signals. Today radio systems are used for applications from walkie-talkie children's toys to the control of space vehicles, as well as for broadcasting, and many other applications.

A *radio receiver* receives its input from an antenna, uses electronic filters to separate a wanted radio signal from all other signals picked up by this antenna, amplifies it to a level suitable for further processing, and finally converts through demodulation and decoding the signal into a form usable for the consumer, such as sound, pictures, digital data, measurement values, navigational positions, etc.

Radio Band

Light comparison		
Name	**Frequency (Hz) (Wavelength)**	**Photon energy (eV)**
Gamma ray	> 30 EHz (0.01 nm)	124 keV - 300+ GeV
X-Ray	30 EHz - 30 PHz (0.01 nm - 10 nm)	124 eV to 120 keV
Ultraviolet	30 PHz - 750 THz (10 nm - 400 nm)	3.1 eV to 124 eV
Visible	750 THz - 428.5 THz (400 nm - 700 nm)	1.7 eV - 3.1 eV
Infrared	428.5 THz - 300 GHz (700 nm - 1 mm)	1.24 meV - 1.7 eV
Microwave	300 GHz - 300 MHz (1 mm - 1 m)	1.24 μeV - 1.24 meV
Radio	300 MHz - 3 kHz (1 m - 100 km)	12.4 feV - 1.24 meV

Radio frequencies occupy the range from a 3 kHz to 300 GHz, although commercially important uses of radio use only a small part of this spectrum. Other types of electromagnetic radiation, with frequencies above the RF range, are infrared, visible light, ultraviolet, X-rays and gamma rays. Since the energy of an individual photon of radio frequency is too low to remove an electron from an atom, radio waves are classified as non-ionizing radiation.

Communication Systems

A *radio communication system* sends signals by radio. Types of radio communication systems deployed depend on technology, standards, regulations, radio spectrum allocation, user requirements, service positioning, and investment.

The radio equipment involved in communication systems includes a transmitter and a receiver, each having an antenna and appropriate terminal equipment such as a microphone at the transmitter and a loudspeaker at the receiver in the case of a voice-communication system.

The power consumed in a transmitting station varies depending on the distance of communication and the transmission conditions. The power received at the receiving station is usually only a tiny fraction of the transmitter's output, since communication depends on receiving the information, not the energy, that was transmitted.

Classical radio communications systems use frequency-division multiplexing (FDM) as a strategy to split up and share the available radio-frequency bandwidth for use by different parties communications concurrently. Modern radio communication systems include those that divide up a radio-frequency band by time-division multiplexing (TDM) and code-division multiplexing (CDM) as alternatives to the classical FDM strategy. These systems offer different tradeoffs in supporting multiple users, beyond the FDM strategy that was ideal for broadcast radio but less so for applications such as mobile telephony.

A radio communication system may send information only one way. For example, in broadcasting a single transmitter sends signals to many receivers. Two stations may take turns sending and receiving, using a single radio frequency; this is called "simplex." By using two radio frequencies, two stations may continuously and concurrently send and receive signals - this is called "duplex" operation.

History

In 1864 James Clerk Maxwell showed mathematically that electromagnetic waves could propagate through free space. The effects of electromagnetic waves (then-unexplained "action at a distance" sparking behavior) were actually observed before and after Maxwell's work by many inventors and experimenters including Luigi Galvani (1791), Peter Samuel Munk (1835), Joseph Henry (1842), Samuel Alfred Varley (1852), Edwin Houston, Elihu Thomson, Thomas Edison (1875) and David Edward Hughes (1878). Edison gave the effect the name "etheric force" and Hughes detected a spark impulse up to 500 yards (460 m) with a portable receiver, but none could identify what caused the phenomenon and it was usually written off as electromagnetic induction. In 1886 Heinrich Rudolf Hertz noticed the same sparking phenomenon and, in published experiments (1887-1888), was able to demonstrate the existence of electromagnetic waves in an experiment confirming Maxwell's theory of electromagnetism. The discovery of these "Hertzian waves" (radio waves) prompted many experiments by physicists. An August 1894 lecture by the British physicist Oliver Lodge, where he transmitted and received "Hertzian waves" at distances up to 50 meters, was followed up a year later with experiments by Indian physicist Jagadish Bose in radio microwave optics and construction of a radio based lightning detector by Russian physicist Alexander Stepanovich Popov. Starting in late 1894, Guglielmo Marconi began pursuing the idea of building a wireless telegraphy system based on Hertzian waves (radio). Marconi gained a patent on the system in 1896 and developed it into a commercial communication system over the next few years.

Early 20th century radio systems transmitted messages by continuous wave code only. Early attempts at developing a system of amplitude modulation for voice and music were demonstrated in 1900 and 1906, but had little success. World War I accelerated the development of radio for military communications, and in this era the first vacuum tubes were applied to radio transmitters and receivers. Electronic amplification was a key development in changing radio from an experimental practice by experts into a home appliance. After the war, commercial radio broadcasting began in the 1920s and became an important mass medium for entertainment and news.

World War II again accelerated development of radio for the wartime purposes of aircraft and land communication, radio navigation and radar. After the war, the experiments in television that had been interrupted were resumed, and it also became an important home entertainment medium.

Uses of Radio

Early uses were maritime, for sending telegraphic messages using Morse code between ships and land. The earliest users included the Japanese Navy scouting the Russian fleet during the Battle of Tsushima in 1905. One of the most memorable uses of marine telegraphy was during the sinking of the RMS *Titanic* in 1912, including communications between operators on the sinking ship and nearby vessels, and communications to shore stations listing the survivors.

Radio was used to pass on orders and communications between armies and navies on both sides in World War I; Germany used radio communications for diplomatic messages once it discovered that its submarine cables had been tapped by the British. The United States passed on President Woodrow Wilson's Fourteen Points to Germany via radio during the war. Broadcasting began from San Jose, California in 1909, and became feasible in the 1920s, with the widespread introduction of radio receivers, particularly in Europe and the United States. Besides broadcasting, point-to-point broadcasting, including telephone messages and relays of radio programs, became widespread in the 1920s and 1930s. Another use of radio in the pre-war years was the development of detection and locating of aircraft and ships by the use of radar (*RA*dio *D*etection *A*nd *R*anging).

Today, radio takes many forms, including wireless networks and mobile communications of all types, as well as radio broadcasting. Before the advent of television, commercial radio broadcasts included not only news and music, but dramas, comedies, variety shows, and many other forms of entertainment (the era from the late 1920s to the mid-1950s is commonly called radio's "Golden Age"). Radio was unique among methods of dramatic presentation in that it used only sound.

Audio

One-way

Bakelite radio at the Bakelite Museum, Orchard Mill, Williton, Somerset, UK.

AM radio uses amplitude modulation, in which the amplitude of the transmitted signal is made proportional to the sound amplitude captured (transduced) by the microphone, while the transmitted frequency remains unchanged. Transmissions are affected by static and interference because lightning and other sources of radio emissions on the same frequency add their amplitudes to the original transmitted amplitude.

A Fisher 500 AM/FM hi-fi receiver from 1959.

In the early part of the 20th century, American AM radio stations broadcast with powers as high as 500 kW, and some could be heard worldwide; these stations' transmitters were commandeered for military use by the US Government during World War II. Currently, the maximum broadcast power for a civilian AM radio station in the United States and Canada is 50 kW, and the majority of stations that emit signals this powerful were grandfathered in. In 1986 KTNN received the last granted 50,000-watt class A license. These 50 kW stations are generally called "clear channel" stations, because within North America each of these stations has exclusive use of its broadcast frequency throughout part or all of the broadcast day.

Bush House, old home of the BBC World Service.

FM broadcast radio sends music and voice with less noise than AM radio. It is often mistakenly thought that FM is higher fidelity than AM, but that is not true. AM is capable of the same audio bandwidth that FM employs. AM receivers typically use narrower filters in the receiver to recover the signal with less noise. AM stereo receivers can reproduce the same audio bandwidth that FM does due to the wider filter used in an AM stereo receiver, but today, AM radios limit the audio bandpass to 3–5 kHz. In frequency modulation, amplitude variation at the microphone causes the transmitter frequency to fluctuate. Because the audio signal modulates the frequency and not the amplitude, an FM signal is not subject to static and interference in the same way as AM signals. Due to its need for a wider bandwidth, FM is transmitted in the Very High Frequency (VHF, 30 MHz to 300 MHz) radio spectrum.

VHF radio waves act more like light, traveling in straight lines; hence the reception range is generally limited to about 50–200 miles (80–322 km). During unusual upper atmospheric conditions, FM signals are occasionally reflected back towards the Earth by the ionosphere, resulting in long distance FM reception. FM receivers are subject to the capture effect, which causes the radio to only receive the strongest signal when multiple signals appear on the same frequency. FM receivers are relatively immune to lightning and spark interference.

High power is useful in penetrating buildings, diffracting around hills, and refracting in the dense atmosphere near the horizon for some distance beyond the horizon. Consequently, 100,000-watt FM stations can regularly be heard up to 100 miles (160 km) away, and farther, 150 miles (240 km), if there are no competing signals.

A few old, "grandfathered" stations do not conform to these power rules. WBCT-FM (93.7) in Grand Rapids, Michigan, US, runs 320,000 watts ERP, and can increase to 500,000 watts ERP by the terms of its original license. Such a huge power level does not usually help to increase range as much as one might expect, because VHF frequencies travel in nearly straight lines over the horizon and off into space. Nevertheless, when there were fewer FM stations competing, this station could be heard near Bloomington, Illinois, US, almost 300 miles (480 km) away.

FM subcarrier services are secondary signals transmitted in a "piggyback" fashion along with the main program. Special receivers are required to utilize these services. Analog channels may contain alternative programming, such as reading services for the blind, background music or stereo sound signals. In some extremely crowded metropolitan areas, the sub-channel program might be an alternate foreign-language radio program for various ethnic groups. Sub-carriers can also transmit digital data, such as station identification, the current song's name, web addresses, or stock quotes. In some countries, FM radios automatically re-tune themselves to the same channel in a different district by using sub-bands.

Two-way

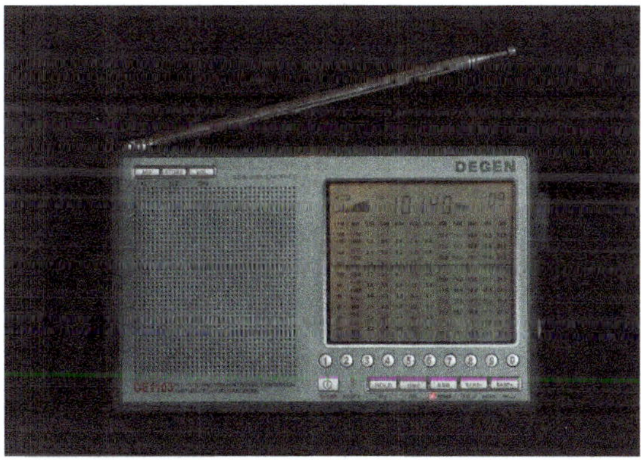

Degen DE1103, an advanced world mini-receiver with single sideband modulation and dual conversion

Aviation voice radios use Aircraft band VHF AM. AM is used so that multiple stations on the same channel can be received. (Use of FM would result in stronger stations blocking out reception of weaker stations due to FM's capture effect). Aircraft fly high enough that their transmitters can be received hundreds of miles away, even though they are using VHF.

Marine voice radios can use single sideband voice (SSB) in the shortwave High Frequency (HF—3 MHz to 30 MHz) radio spectrum for very long ranges or Marine VHF radio / *narrowband FM* in the VHF spectrum for much shorter ranges. Narrowband FM sacrifices fidelity to make more channels available within the radio spectrum, by using a smaller range of radio frequencies, usually with five kHz of deviation, versus the 75 kHz used by commercial FM broadcasts, and 25 kHz used for TV sound.

Government, police, fire and commercial voice services also use narrowband FM on special frequencies. Early police radios used AM receivers to receive one-way dispatches.

Civil and military HF (high frequency) voice services use shortwave radio to contact ships at sea, aircraft and isolated settlements. Most use single sideband voice (SSB), which uses less bandwidth than AM. On an AM radio SSB sounds like ducks quacking, or the adults in a Charlie Brown cartoon. Viewed as a graph of frequency versus power, an AM signal shows power where the frequencies of the voice add and subtract with the main radio frequency. SSB cuts the bandwidth in half by suppressing the carrier and one of the sidebands. This also makes the transmitter about three times more powerful, because it doesn't need to transmit the unused carrier and sideband.

TETRA, Terrestrial Trunked Radio is a digital cell phone system for military, police and ambulances. Commercial services such as XM, WorldSpace and Sirius offer encrypted digital satellite radio.

Telephony

Mobile phones transmit to a local cell site (transmitter/receiver) that ultimately connects to the public switched telephone network (PSTN) through an optic fiber or microwave radio and other network elements. When the mobile phone nears the edge of the cell site's radio coverage area, the central computer switches the phone to a new cell. Cell phones originally used FM, but now most use various digital modulation schemes. Recent developments in Sweden (such as DROPme) allow for the instant downloading of digital material from a radio broadcast (such as a song) to a mobile phone.

Satellite phones use satellites rather than cell towers to communicate.

Video

Analog television sends the picture as AM and the sound as AM or FM, with the sound carrier a fixed frequency (4.5 MHz in the NTSC system) away from the video carrier. Analog television also uses a vestigial sideband on the video carrier to reduce the bandwidth required.

Digital television uses 8VSB modulation in North America (under the ATSC digital television standard), and COFDM modulation elsewhere in the world (using the DVB-T standard). A Reed–Solomon error correction code adds redundant correction codes and allows reliable reception during moderate data loss. Although many current and future codecs can be sent in the MPEG transport stream container format, as of 2006 most systems use a standard-definition format almost identical to DVD: MPEG-2 video in Anamorphic widescreen and MPEG layer 2 (*MP2*) audio. High-definition television is possible simply by using a higher-resolution picture, but H.264/AVC is being considered as a replacement video codec in some regions for its improved compression. With the compression and improved modulation involved, a single "channel" can contain a high-definition program and several standard-definition programs.

Navigation

All satellite navigation systems use satellites with precision clocks. The satellite transmits its position, and the time of the transmission. The receiver listens to four satellites, and can figure its position as being on a line that is tangent to a spherical shell around each satellite, determined by the time-of-flight of the radio signals from the satellite. A computer in the receiver does the math.

Radio direction-finding is the oldest form of radio navigation. Before 1960 navigators used movable loop antennas to locate commercial AM stations near cities. In some cases they used marine radiolocation beacons, which share a range of frequencies just above AM radio with amateur radio operators. LORAN systems also used time-of-flight radio signals, but from radio stations on the ground.

Very High Frequency Omnidirectional Range (VOR), systems (used by aircraft), have an antenna array that transmits two signals simultaneously. A directional signal rotates like a lighthouse at a fixed rate. When the directional signal is facing north, an omnidirectional signal pulses. By measuring the difference in phase of these two signals, an aircraft can determine its bearing or radial from the station, thus establishing a line of position. An aircraft can get readings from two VORs and locate its position at the intersection of the two radials, known as a "fix."

When the VOR station is collocated with DME (Distance Measuring Equipment), the aircraft can determine its bearing and range from the station, thus providing a fix from only one ground station. Such stations are called VOR/DMEs. The military operates a similar system of navaids, called TACANs, which are often built into VOR stations. Such stations are called VORTACs. Because TACANs include distance measuring equipment, VOR/DME and VORTAC stations are identical in navigation potential to civil aircraft.

Radar

Radar (Radio Detection And Ranging) detects objects at a distance by bouncing radio waves off them. The delay caused by the echo measures the distance. The direction of the beam determines the direction of the reflection. The polarization and frequency of the return can sense the type of surface. Navigational radars scan a wide area two to four times per minute. They use very short waves that reflect from earth and stone. They are common on commercial ships and long-distance commercial aircraft.

General purpose radars generally use navigational radar frequencies, but modulate and polarize the pulse so the receiver can determine the type of surface of the reflector. The best general-purpose radars distinguish the rain of heavy storms, as well as land and vehicles. Some can superimpose sonar data and map data from GPS position.

Search radars scan a wide area with pulses of short radio waves. They usually scan the area two to four times a minute. Sometimes search radars use the Doppler effect to separate moving vehicles from clutter. Targeting radars use the same principle as search radar but scan a much smaller area far more often, usually several times a second or more. Weather radars resemble search radars, but use radio waves with circular polarization and a wavelength to reflect from water droplets. Some weather radar use the Doppler effect to measure wind speeds.

Data (Digital Radio)

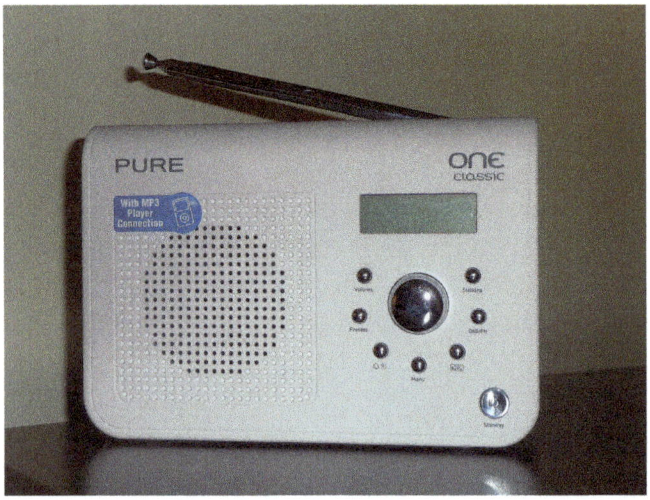

2008 Pure One Classic digital radio

Most new radio systems are digital, including Digital TV, satellite radio, and Digital Audio Broadcasting. The oldest form of digital broadcast was spark gap telegraphy, used by pioneers such as Marconi. By pressing the key, the operator could send messages in Morse code by energizing a rotating commutating spark gap. The rotating commutator produced a tone in the receiver, where a simple spark gap would produce a hiss, indistinguishable from static. Spark-gap transmitters are now illegal, because their transmissions span several hundred megahertz. This is very wasteful of both radio frequencies and power.

The next advance was continuous wave telegraphy, or CW (Continuous Wave), in which a pure radio frequency, produced by a vacuum tube electronic oscillator was switched on and off by a key. A receiver with a local oscillator would “heterodyne” with the pure radio frequency, creating a whistle-like audio tone. CW uses less than 100 Hz of bandwidth. CW is still used, these days primarily by amateur radio operators (hams). Strictly, on-off keying of a carrier should be known as “Interrupted Continuous Wave” or ICW or on-off keying (OOK).

Radioteletype equipment usually operates on short-wave (HF) and is much loved by the military because they create written information without a skilled operator. They send a bit as one of two tones using frequency-shift keying. Groups of five or seven bits become a character printed by a teleprinter. From about 1925 to 1975, radioteletype was how most commercial messages were sent to less developed countries. These are still used by the military and weather services.

Aircraft use a 1200 Baud radioteletype service over VHF to send their ID, altitude and position, and get gate and connecting-flight data. Microwave dishes on satellites, telephone exchanges and

TV stations usually use quadrature amplitude modulation (QAM). QAM sends data by changing both the phase and the amplitude of the radio signal. Engineers like QAM because it packs the most bits into a radio signal when given an exclusive (non-shared) fixed narrowband frequency range. Usually the bits are sent in "frames" that repeat. A special bit pattern is used to locate the beginning of a frame.

Modern GPS receivers.

Communication systems that limit themselves to a fixed narrowband frequency range are vulnerable to jamming. A variety of jamming-resistant spread spectrum techniques were initially developed for military use, most famously for Global Positioning System satellite transmissions. Commercial use of spread spectrum began in the 1980s. Bluetooth, most cell phones, and the 802.11b version of Wi-Fi each use various forms of spread spectrum.

Systems that need reliability, or that share their frequency with other services, may use "coded orthogonal frequency-division multiplexing" or COFDM. COFDM breaks a digital signal into as many as several hundred slower subchannels. The digital signal is often sent as QAM on the subchannels. Modern COFDM systems use a small computer to make and decode the signal with digital signal processing, which is more flexible and far less expensive than older systems that implemented separate electronic channels.

COFDM resists fading and ghosting because the narrow-channel QAM signals can be sent slowly. An adaptive system, or one that sends error-correction codes can also resist interference, because most interference can affect only a few of the QAM channels. COFDM is used for Wi-Fi, some cell phones, Digital Radio Mondiale, Eureka 147, and many other local area network, digital TV and radio standards.

Heating

Radio-frequency energy generated for heating of objects is generally not intended to radiate outside of the generating equipment, to prevent interference with other radio signals. Microwave ovens use intense radio waves to heat food. Diathermy equipment is used in surgery for sealing of blood vessels. Induction furnaces are used for melting metal for casting, and induction hobs for cooking.

Amateur Radio Service

Amateur radio, also known as "ham radio", is a hobby in which enthusiasts are licensed to communicate on a number of bands in the radio frequency spectrum non-commercially and for their own experiments. They may also provide emergency and service assistance in exceptional circumstances. This contribution has been very beneficial in saving lives in many instances.

Amateur radio station with multiple receivers and transceivers

Radio amateurs use a variety of modes, including efficient ones like Morse code and experimental ones like Low-Frequency Experimental Radio. Several forms of radio were pioneered by radio amateurs and later became commercially important, including FM, single-sideband (SSB), AM, digital packet radio and satellite repeaters. Some amateur frequencies may be disrupted illegally by power-line internet service.

Unlicensed Radio Services

Unlicensed, government-authorized personal radio services such as Citizens' band radio in Australia, most of the Americas, and Europe, and Family Radio Service and Multi-Use Radio Service in North America exist to provide simple, usually short range communication for individuals and small groups, without the overhead of licensing. Similar services exist in other parts of the world. These radio services involve the use of handheld units.

Wi-Fi also operates in unlicensed radio bands and is very widely used to network computers.

Free radio stations, sometimes called pirate radio or "clandestine" stations, are unauthorized, unlicensed, illegal broadcasting stations. These are often low power transmitters operated on sporadic schedules by hobbyists, community activists, or political and cultural dissidents. Some pirate stations operating offshore in parts of Europe and the United Kingdom more closely resembled legal stations, maintaining regular schedules, using high power, and selling commercial advertising time.

Radio Control (RC)

Radio remote controls use radio waves to transmit control data to a remote object as in some early forms of guided missile, some early TV remotes and a range of model boats, cars and airplanes. Large industrial remote-controlled equipment such as cranes and switching locomotives now usually use digital radio techniques to ensure safety and reliability.

In Madison Square Garden, at the Electrical Exhibition of 1898, Nikola Tesla successfully demonstrated a radio-controlled boat. He was awarded U.S. patent No. 613,809 for a "Method of and Apparatus for Controlling Mechanism of Moving Vessels or Vehicles."

Radio Propagation

Radio propagation is the behavior of radio waves as they travel, or are propagated, from one point to another, or into various parts of the atmosphere. As a form of electromagnetic radiation, like light waves, radio waves are affected by the phenomena of reflection, refraction, diffraction, absorption, polarization, and scattering.

Radio propagation is affected by the daily changes of water vapor in the troposphere and ionization in the upper atmosphere, due to the Sun. Understanding the effects of varying conditions on radio propagation has many practical applications, from choosing frequencies for international shortwave broadcasters, to designing reliable mobile telephone systems, to radio navigation, to operation of radar systems.

Radio propagation is also affected by several other factors determined by its path from point to point. This path can be a direct line of sight path or an over-the-horizon path aided by refraction in the ionosphere, which is a region between approximately 60 and 600 km above the earth's surface. Factors influencing ionospheric radio signal propagation can include sporadic-E, spread-F, solar flares, geomagnetic storms, ionospheric layer tilts, and solar proton events.

Radio waves at different frequencies propagate in different ways. At extremely low frequencies (ELF) and very low frequencies the wavelength is much larger than the separation between the earth's surface and the D layer of the ionosphere, so electromagnetic waves may propagate in this region as a waveguide. Indeed, for frequencies below 20 kHz, the wave propagates as a single waveguide mode with a horizontal magnetic field and vertical electric field. The interaction of radio waves with the ionized regions of the atmosphere makes radio propagation more complex to predict and analyze than in free space. Ionospheric radio propagation has a strong connection to space weather. A sudden ionospheric disturbance or shortwave fadeout is observed when the x-rays associated with a solar flare ionize the ionospheric D region. Enhanced ionization in that region increases the absorption of radio signals passing through it. During the strongest solar x-ray flares, complete absorption of virtually all ionospherically propagated radio signals in the sunlit hemisphere can occur. These solar flares can disrupt HF radio propagation and affect GPS accuracy.

Since radio propagation is not fully predictable, such services as emergency locator transmitters, in-flight communication with ocean-crossing aircraft, and some television broadcasting have been moved to communications satellites. A satellite link, though expensive, can offer highly predictable and stable line of sight coverage of a given area.

Free Space Propagation

In free space, all electromagnetic waves (radio, light, X-rays, etc.) obey the inverse-square law which states that the power density of an electromagnetic wave is proportional to the inverse of the square of the distance from a point source or:

Doubling the distance from a transmitter means that the power density of the radiated wave at that new location is reduced to one-quarter of its previous value.

The power density per surface unit is proportional to the product of the electric and magnetic field strengths. Thus, doubling the propagation path distance from the transmitter reduces each of their received field strengths over a free-space path by one-half.

Modes

Radio frequencies and their primary mode of propagation

Band		Frequency	Wavelength	Propagation via
ELF	Extremely Low Frequency	3–30 Hz	10,000-100,000 km	
SLF	Super Low Frequency	30–300 Hz	10,000-1,000 km	
ULF	Ultra Low Frequency	0.3–3 kHz (300–3,000 Hz)	1,000–100 km	
VLF	Very Low Frequency	3–30 kHz (3,000–30,000 Hz)	100–10 km	Guided between the earth and the ionosphere.
LF	Low Frequency	30–300 kHz (30,000–300,000 Hz)	10–1 km	Guided between the earth and the D layer of the ionosphere. Surface waves.
MF	Medium Frequency	300–3000 kHz (300,000–3,000,000 Hz)	1000–100 m	Surface waves. E, F layer ionospheric refraction at night, when D layer absorption weakens.
HF	High Frequency (Short Wave)	3–30 MHz (3,000,000–30,000,000 Hz)	100–10 m	E layer ionospheric refraction. F1, F2 layer ionospheric refraction.
VHF	Very High Frequency	30–300 MHz (30,000,000–300,000,000 Hz)	10–1 m	Infrequent E ionospheric (E_s) refraction. Uncommonly F2 layer ionospheric refraction during high sunspot activity up to 50 MHz and rarely to 80 MHz. Generally direct wave. Sometimes tropospheric ducting or meteor scatter

UHF	Ultra High Frequency	300–3000 MHz (300,000,000–3,000,000,000 Hz)	100–10 cm	Direct wave. Sometimes tropospheric ducting.
SHF	Super High Frequency	3–30 GHz (3,000,000,000–30,000,000,000 Hz)	10–1 cm	Direct wave. Sometimes rain scatter.
EHF	Extremely High Frequency	30–300 GHz (30,000,000,000–300,000,000,000 Hz)	10–1 mm	Direct wave limited by absorption.
THF	Tremendously High frequency	0.3–3 THz (300,000,000,000–3,000,000,000,000 Hz)	1–0.1 mm	

Surface Modes (Groundwave)

Lower frequencies (between 30 and 3,000 kHz) have the property of following the curvature of the earth via -[groundwave] propagation in the majority of occurrences.

In this mode the radio wave propagates by interacting with the semi-conductive surface of the earth. The wave "clings" to the surface and thus follows the curvature of the earth. Vertical polarization is used to alleviate short circuiting the electric field through the conductivity of the ground. Since the ground is not a perfect electrical conductor, ground waves are attenuated rapidly as they follow the earth's surface. Attenuation is proportional to the frequency making this mode mainly useful for LF and VLF frequencies.

Today LF and VLF are mostly used for time signals, and for military communications, especially one-way transmissions to ships and submarines, although radio amateurs have an allocation at 137 kHz in some parts of the world. Radio broadcasting using surface wave propagation uses the higher portion of the LF range in Europe, Africa and the Middle East.

Early commercial and professional radio services relied exclusively on long wave, low frequencies and ground-wave propagation. To prevent interference with these services, amateur and experimental transmitters were restricted to the higher (HF) frequencies, felt to be useless since their ground-wave range was limited. Upon discovery of the other propagation modes possible at medium wave and short wave frequencies, the advantages of HF for commercial and military purposes became apparent. Amateur experimentation was then confined only to authorized frequency segments in that range.

Direct Modes (Line-of-sight)

Line-of-sight is the direct propagation of radio waves between antennas that are visible to each other. This is probably the most common of the radio propagation modes at VHF and higher frequencies. Because radio signals can travel through many non-metallic objects, radio can be picked up through walls. This is still line-of-sight propagation. Examples would include propagation between a satellite and a ground antenna or reception of television signals from a local TV transmitter.

Ground plane reflection effects are an important factor in VHF line of sight propagation. The interference between the direct beam line-of-sight and the ground reflected beam often leads to an effective inverse-fourth-power ($1/\text{distance}^4$) law for ground-plane limited radiation. [Need reference to inverse-fourth-power law + ground plane. Drawings may clarify]

Ionospheric Modes (Skywave)

Skywave propagation, also referred to as skip, is any of the modes that rely on refraction of radio waves in the ionosphere, which is made up of one or more ionized layers in the upper atmosphere. F2-layer is the most important ionospheric layer for long-distance, multiple-hop HF propagation, though F1, E, and D-layers also play significant roles. The D-layer, when present during sunlight periods, causes significant amount of signal loss, as does the E-layer whose maximum usable frequency can rise to 4 MHz and above and thus block higher frequency signals from reaching the F2-layer. The layers, or more appropriately "regions", are directly affected by the sun on a daily diurnal cycle, a seasonal cycle and the 11-year sunspot cycle and determine the utility of these modes. During solar maxima, or sunspot highs and peaks, the whole HF range up to 30 MHz can be used usually around the clock and F2 propagation up to 50 MHz is observed frequently depending upon daily solar flux 10.7cm radiation values. During solar minima, or minimum sunspot counts down to zero, propagation of frequencies above 15 MHz is generally unavailable.

Although the claim is commonly made that two-way HF propagation along a given path is reciprocal, that is, if the signal from location A reaches location B at a good strength, the signal from location B will be similar at station A because the same path is traversed in both directions. However, the ionosphere is far too complex and constantly changing to support the reciprocity theorem. The path is never exactly the same in both directions. In brief, conditions at the two terminii of a path generally cause dissimilar polarization shifts, dissimilar splits into ordinary rays and extraordinary or *Pedersen rays* which are erratic and impossibly identical or similar due to variations in ionization density, shifting zenith angles, effects of the earth's magnetic DIPOLE contours, antenna radiation patterns, ground conditions and other variables.

Forecasting of skywave modes is of considerable interest to amateur radio operators and commercial marine and aircraft communications, and also to shortwave broadcasters. Real-time propagation can be assessed by listening for transmissions from specific beacon transmitters.

Meteor Scattering

Meteor scattering relies on reflecting radio waves off the intensely ionized columns of air generated by meteors. While this mode is very short duration, often only from a fraction of second to couple of seconds per event, digital Meteor burst communications allows remote stations to communicate to a station that may be hundreds of miles up to over 1,000 miles (1,600 km) away, without the expense required for a satellite link. This mode is most generally useful on VHF frequencies between 30 and 250 MHz.

Auroral Backscatter

Intense columns of Auroral ionization at 100 km altitudes within the auroral oval backscatter radio waves, perhaps most notably on HF and VHF. Backscatter is angle-sensitive—incident ray vs. magnetic field line of the column must be very close to right-angle. Random motions of electrons spiraling around the field lines create a Doppler-spread that broadens the spectra of the emission to more or less noise-like—depending on how high radio frequency is used. The radio-auroras are observed mostly at high latitudes and rarely extend down to middle latitudes. The occurrence of radio-auroras depends on solar activity (flares, coronal holes, CMEs) and annually the events are more numerous during solar cycle maxima. Radio aurora includes the so-called afternoon radio aurora which produces stronger but more distorted signals and after the Harang-minima, the late-night radio aurora (sub-storming phase) returns with variable signal strength and lesser doppler spread. The propagation range for this predominantly back-scatter mode extends up to about 2000 km in east-west plane, but strongest signals are observed most frequently from the north at nearby sites on same latitudes.

Rarely, a strong radio-aurora is followed by Auroral-E, which resembles both propagation types in some ways.

Sporadic-E Propagation

Sporadic E (Es) propagation can be observed on HF and VHF bands. It must not be confused with ordinary HF E-layer propagation. Sporadic-E at mid-latitudes occurs mostly during summer season, from May to August in the northern hemisphere and from November to February in the southern hemisphere. There is no single cause for this mysterious propagation mode. The reflection takes place in a thin sheet of ionisation around 90 km height. The ionisation patches drift westwards at speeds of few hundred km per hour. There is a weak periodicity noted during the season and typically Es is observed on 1 to 3 successive days and remains absent for a few days to reoccur again. Es do not occur during small hours; the events usually begin at dawn, and there is a peak in the afternoon and a second peak in the evening. Es propagation is usually gone by local midnight.

Observation of radio propagation beacons operating around 28.2 MHz, 50 MHz and 70 MHz, indicates that maximum observed frequency (MOF) for Es is found to be lurking around 30 MHz on most days during the summer season, but sometimes MOF may shoot up to 100 MHz or even more in ten minutes to decline slowly during the next few hours. The peak-phase includes oscillation of MOF with periodicity of approximately 5...10 minutes. The propagation range for Es single-hop is typically 1000 to 2000 km, but with multi-hop, double range is observed. The signals are very strong but also with slow deep fading.

Tropospheric Modes

Tropospheric Scattering

At VHF and higher frequencies, small variations (turbulence) in the density of the atmosphere at a height of around 6 miles (10 km) can scatter some of the normally line-of-sight beam of radio frequency energy back toward the ground, allowing over-the-horizon communication between stations as far as 500 miles (800 km) apart. The military developed the White Alice Communications System covering all of Alaska, using this tropospheric scattering principle.

Tropospheric Ducting

Sudden changes in the atmosphere's vertical moisture content and temperature profiles can on random occasions make microwave and UHF & VHF signals propagate hundreds of kilometers up to about 2,000 kilometers (1,300 mi)—and for ducting mode even farther—beyond the normal radio-horizon. The inversion layer is mostly observed over high pressure regions, but there are several tropospheric weather conditions which create these randomly occurring propagation modes. Inversion layer's altitude for non-ducting is typically found between 100 meters (300 ft) to about 1 kilometer (3,000 ft) and for ducting about 500 meters to 3 kilometers (1,600 to 10,000 ft), and the duration of the events are typically from several hours up to several days. Higher frequencies experience the most dramatic increase of signal strengths, while on low-VHF and HF the effect is negligible. Propagation path attenuation may be below free-space loss. Some of the lesser inversion types related to warm ground and cooler air moisture content occur regularly at certain times of the year and time of day. A typical example could be the late summer, early morning tropospheric enhancements that bring in signals from distances up to few hundred kilometers for a couple of hours, until undone by the Sun's warming effect.

Tropospheric Delay

This is a source of error in radio ranging techniques, such as the Global Positioning System (GPS).

Rain Scattering

Rain scattering is purely a microwave propagation mode and is best observed around 10 GHz, but extends down to a few gigahertz—the limit being the size of the scattering particle size vs. wavelength. This mode scatters signals mostly forwards and backwards when using horizontal polarization and side-scattering with vertical polarization. Forward-scattering typically yields propagation ranges of 800 km. Scattering from snowflakes and ice pellets also occurs, but scattering from ice without watery surface is less effective. The most common application for this phenomenon is microwave rain radar, but rain scatter propagation can be a nuisance causing unwanted signals to intermittently propagate where they are not anticipated or desired. Similar reflections may also occur from insects though at lower altitudes and shorter range. Rain also causes attenuation of point-to-point and satellite microwave links. Attenuation values up to 30 dB have been observed on 30 GHz during heavy tropical rain.

Airplane Scattering

Airplane scattering (or most often reflection) is observed on VHF through microwaves and, besides back-scattering, yields momentary propagation up to 500 km even in mountainous terrain. The most common back-scatter applications are air-traffic radar, bistatic forward-scatter guided-missile and airplane-detecting trip-wire radar, and the US space radar.

Lightning Scattering

Lightning scattering has sometimes been observed on VHF and UHF over distances of about 500 km. The hot lightning channel scatters radio-waves for a fraction of a second. The RF noise burst from the lightning makes the initial part of the open channel unusable and the ionization disappears quickly because of recombination at low altitude and high atmospheric pressure. Although the hot lightning channel is briefly observable with microwave radar, no practical use for this mode has been found in communications.

Other Effects

Diffraction

Knife-edge diffraction is the propagation mode where radio waves are bent around sharp edges. For example, this mode is used to send radio signals over a mountain range when a line-of-sight path is not available. However, the angle cannot be too sharp or the signal will not diffract. The diffraction mode requires increased signal strength, so higher power or better antennas will be needed than for an equivalent line-of-sight path.

Diffraction depends on the relationship between the wavelength and the size of the obstacle. In other words, the size of the obstacle in wavelengths. Lower frequencies diffract around large smooth obstacles such as hills more easily. For example, in many cases where VHF (or higher frequency) communication is not possible due to shadowing by a hill, it is still possible to communicate using the upper part of the HF band where the surface wave is of little use.

Diffraction phenomena by small obstacles are also important at high frequencies. Signals for urban cellular telephony tend to be dominated by ground-plane effects as they travel over the rooftops of the urban environment. They then diffract over roof edges into the street, where multipath propagation, absorption and diffraction phenomena dominate.

Absorption

Low-frequency radio waves travel easily through brick and stone and VLF even penetrates sea-water. As the frequency rises, absorption effects become more important. At microwave or higher frequencies, absorption by molecular resonances in the atmosphere (mostly from water, H_2O and oxygen, O_2) is a major factor in radio propagation. For example, in the 58–60 GHz band, there is a major absorption peak which makes this band useless for long-distance use. This phenomenon was first discovered during radar research in World War II. Above about 400 GHz, the Earth's atmosphere blocks most of the spectrum while still passing some - up to UV light, which is blocked by ozone - but visible light and some of the near-infrared is transmitted. Heavy rain and falling snow also affect microwave absorption.

Measuring HF Propagation

HF propagation conditions can be simulated using radio propagation models, such as the Voice of America Coverage Analysis Program, and realtime measurements can be done using chirp transmitters. For radio amateurs the WSPR mode provides maps with real time propagation conditions between a network of transmitters and receivers. Even without special beacons the realtime propagation conditions can be measured: a worldwide network of receivers decodes morse code signals on amateur radio frequencies in realtime and provides sophisticated search functions and propagation maps for every station received.

Practical Effects

The average person can notice the effects of changes in radio propagation in several ways.

In AM broadcasting, the dramatic ionospheric changes that occur overnight in the mediumwave band drive a unique broadcast license scheme, with entirely different transmitter power output levels and directional antenna patterns to cope with skywave propagation at night. Very few stations are allowed to run without modifications during dark hours, typically only those on clear channels in North America. Many stations have no authorization to run at all outside of daylight hours. Otherwise, there would be nothing but interference on the entire broadcast band from dusk until dawn without these modifications.

For FM broadcasting (and the few remaining low-band TV stations), weather is the primary cause for changes in VHF propagation, along with some diurnal changes when the sky is mostly without cloud cover. These changes are most obvious during temperature inversions, such as in the late-night and early-morning hours when it is clear, allowing the ground and the air near it to cool more rapidly. This not only causes dew, frost, or fog, but also causes a slight "drag" on the bottom of the radio waves, bending the signals down such that they can follow the Earth's curvature over the normal radio horizon. The result is typically several stations being heard from another media market — usually a neighboring one, but sometimes ones from a few hundred kilometers away. Ice storms are also the result of inversions, but these normally cause more scattered omnidirection propagation, resulting mainly in interference, often among weather radio stations. In late spring and early summer, a combination of other atmospheric factors can occasionally cause skips that duct high-power signals to places well over 1000km away.

Non-broadcast signals are also affected. Mobile phone signals are in the UHF band, ranging from 700 to over 2600 Megahertz, a range which makes them even more prone to weather-induced propagation changes. In urban (and to some extent suburban) areas with a high population density, this is partly offset by the use of smaller cells, which use lower effective radiated power and beam tilt to reduce interference, and therefore increase frequency reuse and user capacity. However, since this would not be very cost-effective in more rural areas, these cells are larger and so more likely to cause interference over longer distances when propagation conditions allow.

While this is generally transparent to the user thanks to the way that cellular networks handle cell-to-cell handoffs, when cross-border signals are involved, unexepected charges for international roaming may occur despite not having left the country at all. This often occurs between southern San Diego and northern Tijuana at the western end of the U.S./Mexico border, and between eastern Detroit and western Windsor along the U.S./Canada border. Since signals can travel unobstructed over a body of water far larger than the Detroit River, and cool water temperatures also cause inversions in surface air, this "fringe roaming" sometimes occurs across the Great Lakes, and between islands in the Caribbean. Signals can skip from the Dominican Republic to a mountainside in Puerto Rico and vice versa, or between the U.S. and British Virgin Islands, among others. While unintended cross-border roaming is often automatically removed by mobile phone company billing systems, inter-island roaming is typically not.

Radio Frequency

Radio frequency (RF) is any of the electromagnetic wave frequencies that lie in the range extending from around 3 kHz to 300 GHz, which include those frequencies used for communications or radar signals. RF usually refers to electrical rather than mechanical oscillations. However, mechanical RF systems do exist.

Although radio *frequency* is a rate of oscillation, the term "radio frequency" or its abbreviation "RF" are used as a synonym for radio – i.e., to describe the use of wireless communication, as opposed to communication via electric wires. Examples include:

- Radio-frequency identification
- ISO/IEC 14443-2 *Radio frequency power and signal interface*

Special Properties of RF Current

Electric currents that oscillate at radio frequencies have special properties not shared by direct current or alternating current of lower frequencies.

- The energy in an RF current can radiate off a conductor into space as electromagnetic waves (radio waves); this is the basis of radio technology.
- RF current does not penetrate deeply into electrical conductors but tends to flow along their surfaces; this is known as the skin effect. For this reason, when the human body comes in contact with high power RF currents it can cause superficial but serious burns called *RF burns (Note that RF burns result from electrical energy, while Radiation burns result from electromagnetic energy).*
- RF currents applied to the body often do not cause the painful sensation of electric shock as do lower frequency currents. This is because the current changes direction too quickly to trigger depolarization of nerve membranes.
- RF current can easily ionize air, creating a conductive path through it. This property is exploited by "high frequency" units used in electric arc welding, which use currents at higher frequencies than power distribution uses.
- Another property is the ability to appear to flow through paths that contain insulating material, like the dielectric insulator of a capacitor.
- When conducted by an ordinary electric cable, RF current has a tendency to reflect from discontinuities in the cable such as connectors and travel back down the cable toward the source, causing a condition called standing waves. Therefore, RF current must be carried by specialized types of cable called transmission line.

Radio Communication

To receive radio signals an antenna must be used. However, since the antenna will pick up thousands of radio signals at a time, a radio tuner is necessary to *tune into* a particular frequency (or frequency range). This is typically done via a resonator – in its simplest form, a circuit with a capacitor and an inductor form a tuned circuit. The resonator amplifies oscillations within a particular frequency band, while reducing oscillations at other frequencies outside the band. Another method to isolate a particular radio frequency is by oversampling (which gets a wide range of frequencies) and picking out the frequencies of interest, as done in software defined radio.

The distance over which radio communications is useful depends significantly on things other than wavelength, such as transmitter power, receiver quality, type, size, and height of antenna, mode of transmission, noise, and interfering signals. Ground waves, tropospheric scatter and skywaves can all achieve greater ranges than line-of-sight propagation. The study of radio propagation allows estimates of useful range to be made.

Frequency Bands

Frequency	Wavelength	Designation	Abbreviation
3–30 Hz	10^5–10^4 km	Extremely low frequency	ELF
30–300 Hz	10^4–10^3 km	Super low frequency	SLF
300–3000 Hz	10^3–100 km	Ultra low frequency	ULF
3–30 kHz	100–10 km	Very low frequency	VLF
30–300 kHz	10–1 km	Low frequency	LF
300 kHz – 3 MHz	1 km – 100 m	Medium frequency	MF
3–30 MHz	100–10 m	High frequency	HF
30–300 MHz	10–1 m	Very high frequency	VHF
300 MHz – 3 GHz	1 m – 10 cm	Ultra high frequency	UHF
3–30 GHz	10–1 cm	Super high frequency	SHF
30–300 GHz	1 cm – 1 mm	Extremely high frequency	EHF
300 GHz – 3 THz	1 mm – 0.1 mm	Tremendously high frequency	THF

In Medicine

Radio frequency (RF) energy, in the form of radiating waves or electrical currents, has been used in medical treatments for over 75 years, generally for minimally invasive surgeries using radiofrequency ablation including the treatment of sleep apnea. Magnetic resonance imaging (MRI) uses radio frequency waves to generate images of the human body.

Radio frequencies at non-ablation energy levels are sometimes used as a form of cosmetic treatment that can tighten skin, reduce fat (lipolysis), or promote healing.

RF diathermy is a medical treatment that uses RF induced heat as a form of physical or occupational therapy and in surgical procedures. It is commonly used for muscle relaxation. It is also a method of heating tissue electromagnetically for therapeutic purposes in medicine. Diathermy is used in physical therapy and occupational therapy to deliver moderate heat directly to pathologic lesions in the deeper tissues of the body. Surgically, the extreme heat that can be produced by diathermy may be used to destroy neoplasms, warts, and infected tissues, and to cauterize blood vessels to prevent excessive bleeding. The technique is particularly valuable in neurosurgery and surgery of the eye. Diathermy equipment typically operates in the short-wave radio frequency (range 1–100 MHz) or microwave energy (range 434–915 MHz).

Pulsed electromagnetic field therapy (PEMF) is a medical treatment that purportedly helps to heal bone tissue reported in a recent NASA study. This method usually employs electromagnetic radiation of different frequencies - ranging from static magnetic fields, through extremely low frequencies (ELF) to higher radio frequencies (RF) administered in pulses.

Effects on the Human Body

Extremely Low Frequency RF

High-power extremely low frequency RF with electric field levels in the low kV/m range are known to induce perceivable currents within the human body that create an annoying tingling sensation. These currents will typically flow to ground through a body contact surface such as the feet, or arc to ground where the body is well insulated.

Microwaves

Microwave exposure at low-power levels below the Specific absorption rate set by government regulatory bodies are considered harmless non-ionizing radiation and have no effect on the human body. However, levels above the Specific absorption rate set by the U.S. Federal Communications Commission are considered potentially harmful.

Long-term human exposure to high-levels of microwaves is recognized to cause cataracts according to experimental animal studies and epidemiological studies. The mechanism is unclear but may include changes in heat sensitive enzymes that normally protect cell proteins in the lens. Another mechanism that has been advanced is direct damage to the lens from pressure waves induced in the aqueous humor.

High-power exposure to microwave RF is known to create a range of effects from lower to higher power levels, ranging from unpleasant burning sensation on the skin and microwave auditory effect, to extreme pain at the mid-range, to physical burning and blistering of skin and internals at high power levels.

General RF Exposure

The 1999 revision of Canadian Safety Code 6 recommended electric field limits of 100 kV/m for pulsed EMF to prevent air breakdown and spark discharges, mentioning rationale related to auditory effect and energy-induced unconsciousness in rats. The pulsed EMF limit was removed in later revisions, however.

As a Weapon

A heat ray is an RF harassment device that makes use of microwave radio frequencies to create an unpleasant heating effect in the upper layer of the skin. A publicly known heat ray weapon called the Active Denial System was developed by the US military as an experimental weapon to deny the enemy access to an area. A death ray is a weapon that delivers heat ray electromagnetic energy at levels that injure human tissue. The inventor of the death ray, Harry Grindell Matthews, claims to have lost sight in his left eye while developing his death ray weapon based on a primitive microwave magnetron from the 1920s (note that a typical microwave oven induces a tissue damaging cooking effect inside the oven at about 2 kV/m.)

Measurement

Since radio frequency radiation has both an electric and a magnetic component, it is often convenient to express intensity of radiation field in terms of units specific to each component. The unit *volts per meter* (V/m) is used for the electric component, and the unit *amperes per meter* (A/m) is used for the magnetic component. One can speak of an electromagnetic field, and these units are used to provide information about the levels of electric and magnetic field strength at a measurement location.

Another commonly used unit for characterizing an RF electromagnetic field is *power density*. Power density is most accurately used when the point of measurement is far enough away from the RF emitter to be located in what is referred to as the far field zone of the radiation pattern. In closer proximity to the transmitter, i.e., in the "near field" zone, the physical relationships between the electric and magnetic components of the field can be complex, and it is best to use the field strength units discussed above. Power density is measured in terms of power per unit area, for example, milliwatts per square centimeter (mW/cm^2). When speaking of frequencies in the microwave range and higher, power density is usually used to express intensity since exposures that might occur would likely be in the far field zone.

Skywave

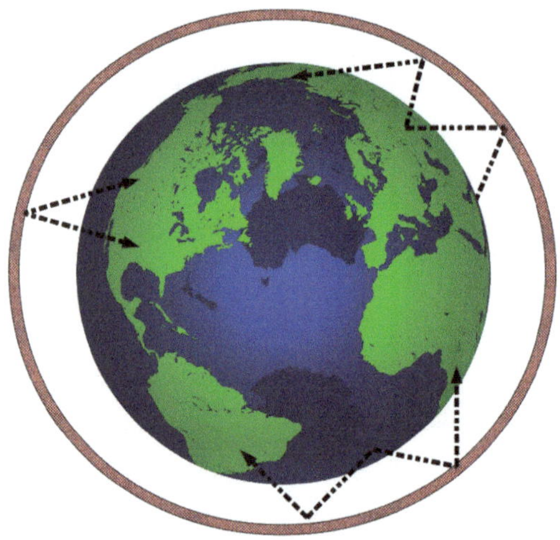

Radio waves (black) reflecting off the ionosphere (red) during skywave propagation.

In radio communication, skywave or skip refers to the propagation of radio waves reflected or refracted back toward Earth from the ionosphere, an electrically charged layer of the upper atmosphere. Since it is not limited by the curvature of the Earth, skywave propagation can be used to communicate beyond the horizon, at intercontinental distances. It is mostly used in the shortwave frequency bands.

As a result of skywave propagation, a signal from a distant AM broadcasting station, a shortwave station, or—during sporadic E propagation conditions (principally during the summer months in both hemispheres)—a low frequency television station can sometimes be received as clearly as local stations. Most long-distance shortwave (high frequency) radio communication—between 3 and 30 MHz—is a result of skywave propagation. Since the early 1920s amateur radio operators (or "hams"), limited to lower transmitter power than broadcast stations, have taken advantage of skywave for long distance (or "DX") communication.

Skywave propagation is distinct from:

- groundwave propagation, where radio waves travel near Earth's surface without being reflected or refracted by the atmosphere—the dominant propagation mode at lower frequencies,
- line-of-sight propagation, in which radio waves travel in a straight line, the dominant mode at higher frequencies.

Explanation

The ionosphere is a region of the upper atmosphere, from about 80 km to 1000 km in altitude, where neutral air is ionized by solar photons and cosmic rays. When high frequency signals enter the ionosphere obliquely, they are back-scattered from the ionized layer as scatter waves. If the midlayer ionization is strong enough compared to the signal frequency, a scatter wave can exit the bottom of the layer earthwards as if reflected from a mirror. Earth's surface (ground or water) then diffusely reflects the incoming wave back towards the ionosphere. Consequently, like a rock "skipping" across water, the signal may effectively "bounce" or "skip" between the earth and ionosphere two or more times (multihop propagation). Since at shallow incidence losses remain quite small, signals of only a few watts can sometimes be received many thousands of miles away as a result. This is what enables shortwave broadcasts to travel all over the world.

If the ionization is not great enough, the scatter wave is initially deflected downwards, and subsequently upwards (above the layer peak) such that it exits the top of the layer slightly displaced. Sky wave propagation occurs in the waveguide formed by the ground and ionosphere, each serving as reflectors. With a single "hop," path distances up to 3500 km may be reached. Transatlantic connections are mostly obtained with two or three hops.

The layer of ionospheric plasma with equal ionization (the reflective surface) is not fixed, but undulates like the surface of the ocean. Varying reflection efficiency from this changing surface can cause the reflected signal strength to change, causing "*fading*" in shortwave broadcasts.

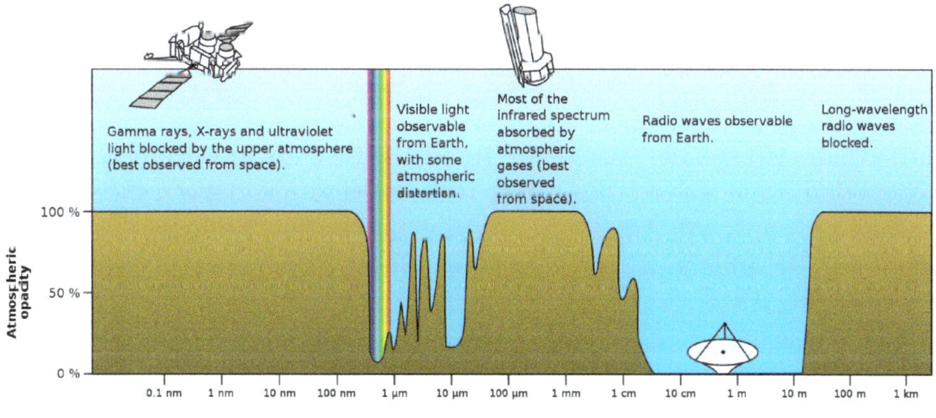

Rough plot of Earth's atmospheric transmittance (or opacity) to various wavelengths of electromagnetic radiation, including radio waves.

Depending on the transmitting antenna, signals below approximately 10 MHz during the day and 5 MHz at night, entering the ionosphere at a steep angle (near-vertical incidence) may be back-scattered down to Earth within a short range. Alternatively, signals beamed close to the horizon enter the ionosphere at a shallow angle and return to Earth over medium to long distances.

Other Considerations

VHF signals with frequencies above about 30 MHz usually penetrate the ionosphere and are not returned to the Earth's surface. E-skip is a notable exception, where VHF signals including FM broadcast and VHF TV signals are frequently reflected to the Earth during late Spring and early Summer. E-skip rarely affects UHF frequencies, except for very rare occurrences below 500 MHz.

Frequencies below approximately 10 MHz (wavelengths longer than 30 meters), including broadcasts in the mediumwave and shortwave bands (and to some extent longwave), propagate most efficiently by skywave at night. Frequencies above 10 MHz (wavelengths shorter than 30 meters) typically propagate most efficiently during the day. Frequencies lower than 3 kHz have a wavelength longer than the distance between the Earth and the ionosphere. The maximum usable frequency for skywave propagation is strongly influenced by sunspot number.

Skywave propagation is usually degraded—sometimes seriously—during geomagnetic storms. Skywave propagation on the sunlit side of the Earth can be entirely disrupted during sudden ionospheric disturbances.

Because the lower-altitude layers (the E-layer in particular) of the ionosphere largely disappear at night, the refractive layer of the ionosphere is much higher above the surface of the Earth at night. This leads to an increase in the "skip" or "hop" distance of the skywave at night.

History

Discovery of Skywave Propagation

Amateur radio operators are credited with the discovery of skywave propagation on the shortwave bands. Early long-distance services used surface wave propagation at very low frequencies, which are attenuated along the path. Longer distances and higher frequencies using this method meant more signal attenuation. This, and the difficulties of generating and detecting higher frequencies, made discovery of shortwave propagation difficult for commercial services.

Radio amateurs conducted the first successful transatlantic tests in December 1921, operating in the 200 meter mediumwave band (1500 kHz)—the shortest wavelength then available to amateurs. In 1922 hundreds of North American amateurs were heard in Europe at 200 meters and at least 30 North American amateurs heard amateur signals from Europe. The first two-way communications between North American and Hawaiian amateurs began in 1922 at 200 meters. Although operation on wavelengths shorter than 200 meters was technically illegal (but tolerated as the authorities mistakenly believed at first that such frequencies were useless for commercial or military use), amateurs began to experiment with those wavelengths using newly available vacuum tubes shortly after World War I.

Extreme interference at the upper edge of the 150-200 meter band—the official wavelengths allocated to amateurs by the Second National Radio Conference in 1923—forced amateurs to shift to shorter and shorter wavelengths; however, amateurs were limited by regulation to wavelengths longer than 150 meters (2 MHz). A few fortunate amateurs who obtained special permission for experimental communications below 150 meters completed hundreds of long distance two way contacts on 100 meters (3 MHz) in 1923 including the first transatlantic two way contacts in November 1923, on 110 meters (2.72 MHz)

By 1924 many additional specially licensed amateurs were routinely making transoceanic contacts at distances of 6000 miles (~9600 km) and more. On 21 September several amateurs in California completed two way contacts with an amateur in New Zealand. On 19 October amateurs in New Zealand and England completed a 90-minute two-way contact nearly halfway around the world. On October 10, the Third National Radio Conference made three shortwave bands available to U.S. amateurs at 80 meters (3.75 MHz), 40 meters (7 MHz) and 20 meters (14 MHz). These were allocated worldwide, while the 10-meter band (28 MHz) was created by the Washington International Radiotelegraph Conference on 25 November 1927. The 15-meter band (21 MHz) was opened to amateurs in the United States on 1 May 1952.

Marconi

In June and July 1923, Guglielmo Marconi's transmissions were completed during nights on 97 meters from Poldhu Wireless Station, Cornwall, to his yacht Ellette in the Cape Verde Islands. In September 1924, Marconi transmitted during daytime and nighttime on 32 meters from Poldhu to his yacht in Beirut. Marconi, in July 1924, entered into contracts with the British General Post Office (GPO) to install high speed shortwave telegraphy circuits from London to Australia, India, South Africa and Canada as the main element of the Imperial Wireless Chain. The UK-to-Canada shortwave "Beam Wireless Service" went into commercial operation on 25 October 1926. Beam Wireless Services from the UK to Australia, South Africa and India went into service in 1927.

Far more spectrum is available for long distance communication in the shortwave bands than in the long wave bands; and shortwave transmitters, receivers and antennas were orders of magnitude less expensive than the multi-hundred kilowatt transmitters and monstrous antennas needed for long wave.

Shortwave communications began to grow rapidly in the 1920s, similar to the internet in the late 20th century. By 1928, more than half of long distance communications had moved from transoceanic cables and long wave wireless services to shortwave "skip" transmission and the overall volume of transoceanic shortwave communications had vastly increased. Shortwave also ended the need for multimillion-dollar investments in new transoceanic telegraph cables and massive long wave wireless stations, although some existing transoceanic telegraph cables and commercial long wave communications stations remained in use until the 1960s.

The cable companies began to lose large sums of money in 1927, and a serious financial crisis threatened the viability of cable companies that were vital to strategic British interests. The British government convened the Imperial Wireless and Cable Conference in 1928 "to examine the situation that had arisen as a result of the competition of Beam Wireless with the Cable Services". It recommended and received Government approval for all overseas cable and wireless resources of the Empire to be merged into one system controlled by a newly formed company in 1929, Imperial and International Communications Ltd. The name of the company was changed to Cable and Wireless Ltd. in 1934.

Radio Broadcasting

Long wave radio broadcasting station, Motala, Sweden

Slovak Radio Building, Bratislava, Slovakia
(architects: Štefan Svetko, Štefan Ďurkovič and Barnabáš Kissling, 1967-1983)

Broadcasting tower in Trondheim, Norway

Radio broadcasting is a unidirectional wireless transmission over radio waves intended to reach a wide audience. Stations can be linked in radio networks to broadcast a common radio format, either in broadcast syndication or simulcast or both. Audio broadcasting also can be done via cable radio, local wire television networks, satellite radio, and internet radio via streaming media on the Internet. The signal types can be either analog audio or digital audio.

History

The earliest radio stations were simply radiotelegraphy systems and did not carry audio. For audio broadcasts to be possible electronic detection and amplification devices had to be incorporated.

The thermionic valve was invented in 1904 by the English physicist John Ambrose Fleming. He developed a device he called an "oscillation valve" (because it passes current in only one direction). The heated filament, or cathode, was capable of thermionic emission of electrons that would flow to the *plate* (or *anode*) when it was at a higher voltage. Electrons, however, could not pass in the reverse direction because the plate was not heated and thus not capable of thermionic emission of electrons. Later known as the Fleming valve, it could be used as a rectifier of alternating current and as a radio wave detector. This greatly improved the crystal set which rectified the radio signal using an early solid-state diode based on a crystal and a so-called cat's whisker. However, what was still required was an amplifier.

The triode (mercury-vapor filled with a control grid) was patented on March 4, 1906 by the Austrian Robert von Lieben independent from that, on October 25, 1906 Lee De Forest patented his three-element Audion. It wasn't put to practical use until 1912, when its amplifying ability became recognized by researchers.

By about 1920, valve technology had matured to the point where radio broadcasting was quickly becoming viable. However, an early audio transmission that could be termed a *broadcast* may have occurred on Christmas Eve in 1906 by Reginald Fessenden, although this is disputed. While many early experimenters attempted to create systems similar to radiotelephone devices by which only two parties were meant to communicate, there were others who intended to transmit to larger audiences. Charles Herrold started broadcasting in California in 1909 and was carrying audio by the next year. (Herrold's station eventually became KCBS).

In The Hague, the Netherlands, PCGG started broadcasting on November 6, 1919, making it, arguably the first commercial broadcasting station. In 1916, Frank Conrad, an electrical engineer employed at the Westinghouse Electric Corporation, began broadcasting from his Wilkinsburg, Pennsylvania garage with the call letters 8XK. Later, the station was moved to the top of the Westinghouse factory building in East Pittsburgh, Pennsylvania. Westinghouse relaunched the station as KDKA on November 2, 1920, as the first commercially licensed radio station in America. The commercial broadcasting designation came from the type of broadcast license; advertisements did not air until years later. The first licensed broadcast in the United States came from KDKA itself: the results of the Harding/Cox Presidential Election. The Montreal station that became CFCF began broadcast programming on May 20, 1920, and the Detroit station that became WWJ began program broadcasts beginning on August 20, 1920, although neither held a license at the time.

In 1920 wireless broadcasts for entertainment began in the UK from the Marconi Research Centre 2MT at Writtle near Chelmsford, England. A famous broadcast from Marconi's New Street Works factory in Chelmsford was made by the famous soprano Dame Nellie Melba on 15 June 1920, where she sang two arias and her famous trill. She was the first artist of international renown to participate in direct radio broadcasts. The 2MT station began to broadcast regular entertainment in 1922. The BBC was amalgamated in 1922 and received a Royal Charter in 1926, making it the first national broadcaster in the world, followed by Czech Radio and other European broadcasters in 1923.

Radio Argentina began regularly scheduled transmissions from the Teatro Coliseo in Buenos Aires on August 27, 1920, making its own priority claim. The station got its license on November 19, 1923. The delay was due to the lack of official Argentine licensing procedures before that date. This station continued regular broadcasting of entertainment and cultural fare for several decades.

Radio in education soon followed and colleges across the U.S. began adding radio broadcasting courses to their curricula. Curry College in Milton, Massachusetts introduced one of the first broadcasting majors in 1932 when the college teamed up with WLOE in Boston to have students broadcast programs.

Types

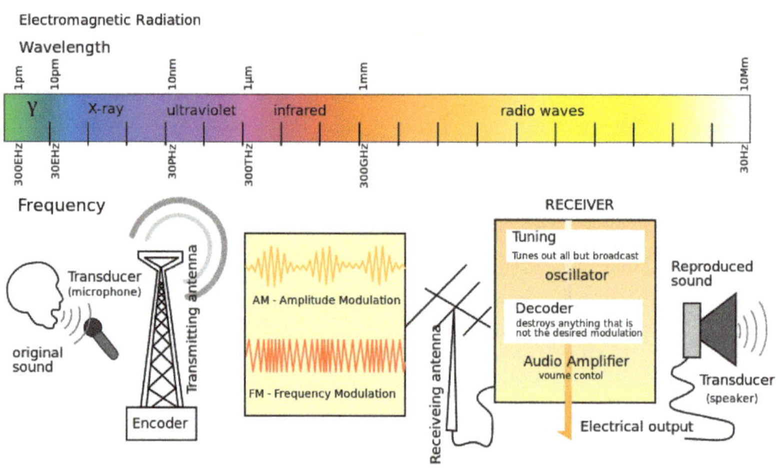

Transmission diagram of sound broadcasting (AM and FM)

Broadcasting by radio takes several forms. These include AM and FM stations. There are several subtypes, namely commercial broadcasting, non-commercial educational (NCE) public broadcasting and non-profit varieties as well as community radio, student-run campus radio stations and hospital radio stations can be found throughout the world.

Many stations broadcast on shortwave bands using AM technology that can be received over thousands of miles (especially at night). For example, the BBC, VOA, VOR, and Deutsche Welle have transmitted via shortwave to Africa and Asia. These broadcasts are very sensitive to atmospheric conditions and solar activity.

Nielsen Audio, formerly known as Arbitron, the United States-based company that reports on radio audiences, defines a "radio station" as a government-licensed AM or FM station; an HD Radio (primary or multicast) station; an internet stream of an existing government-licensed station; one of the satellite radio channels from XM Satellite Radio or Sirius Satellite Radio; or, potentially, a station that is not government licensed.

Shortwave

See shortwave for the differences between shortwave, medium wave and long wave spectra. Shortwave is used largely for national broadcasters, international propaganda, or religious broadcasting organizations.

AM

AM stations were the earliest broadcasting stations to be developed. AM refers to amplitude modulation, a mode of broadcasting radio waves by varying the amplitude of the carrier signal in response to the amplitude of the signal to be transmitted.

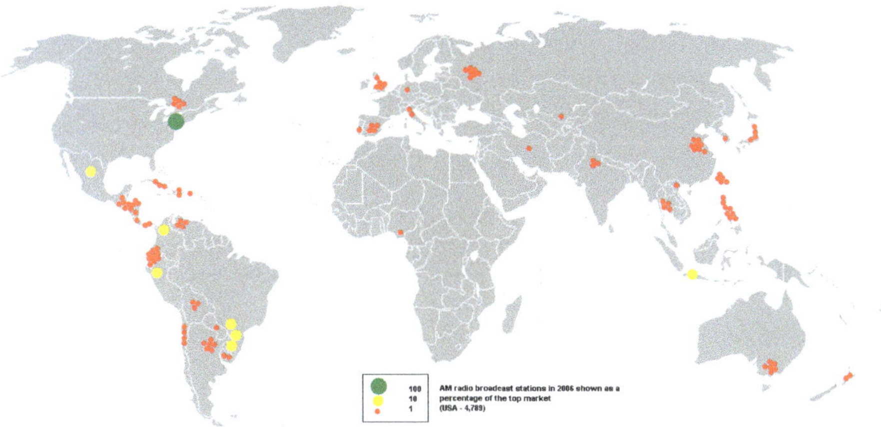

AM broadcasting stations in 2006

The medium-wave band is used worldwide for AM broadcasting. Europe also uses the long wave band. In response to the growing popularity of FM stereo radio stations in the late 1980s and early 1990s, some North American stations began broadcasting in AM stereo, though this never gained popularity, and very few receivers were ever sold.

One of the advantages of AM is that its signal can be detected (turned into sound) with simple equipment. If a signal is strong enough, not even a power source is needed; building an unpowered crystal radio receiver was a common childhood project in the early decades of AM broadcasting.

AM broadcasts occur on North American airwaves in the medium wave frequency range of 530 to 1700 kHz (known as the "standard broadcast band"). The band was expanded in the 1990s by adding nine channels from 1620 to 1700 kHz. Channels are spaced every 10 kHz in the Americas, and generally every 9 kHz everywhere else.

The signal is subject to interference from electrical storms (lightning) and other electromagnetic interference (EMI).

AM transmissions cannot be ionospherically propagated during the day due to strong absorption in the D-layer of the ionosphere. In a crowded channel environment this means that the power of regional channels which share a frequency must be reduced at night or directionally beamed in order to avoid interference, which reduces the potential nighttime audience. Some stations have frequencies unshared with other stations in North America; these are called clear-channel stations. Many of them can be heard across much of the country at night. During the night, absorption largely disappears and permits signals to travel to much more distant locations via ionospheric reflections. However, fading of the signal can be severe at night.

AM radio transmitters can transmit audio frequencies up to 15 kHz (now limited to 10 kHz in the US due to FCC rules designed to reduce interference), but most receivers are only capable of reproducing frequencies up to 5 kHz or less. At the time that AM broadcasting began in the 1920s, this provided adequate fidelity for existing microphones, 78 rpm recordings, and loudspeakers. The fidelity of sound equipment subsequently improved considerably, but the receivers did not. Reducing the bandwidth of the receivers reduces the cost of manufacturing and makes them less prone to interference. AM stations are never assigned adjacent channels in the same service area. This prevents the sideband power generated by two stations from interfering with each other. Bob Carver created an AM stereo tuner employing notch filtering that demonstrated that an AM broadcast can meet or exceed the 15 kHz baseband bandwidth allotted to FM stations without objectionable interference. After several years, the tuner was discontinued. Bob Carver had left the company and the Carver Corporation later cut the number of models produced before discontinuing production completely.

FM

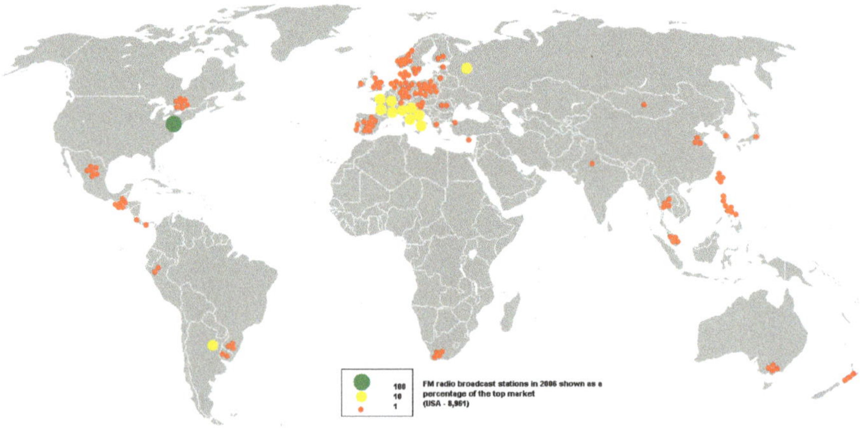

FM radio broadcast stations in 2006

FM refers to frequency modulation, and occurs on VHF airwaves in the frequency range of 88 to 108 MHz everywhere except Japan and Russia. Russia, like the former Soviet Union, uses 65.9 to 74 MHz frequencies in addition to the world standard. Japan uses the 76 to 90 MHz frequency band.

Edwin Howard Armstrong invented FM radio to overcome the problem of radio-frequency interference (RFI), which plagued AM radio reception. At the same time, greater fidelity was made possible by spacing stations further apart in the radio frequency spectrum. Instead of 10 kHz apart, as on the AM band in the US, FM channels are 200 kHz (0.2 MHz) apart. In other countries greater spacing is sometimes mandatory, such as in New Zealand, which uses 700 kHz spacing (previously 800 kHz). The improved fidelity made available was far in advance of the audio equipment of the 1940s, but wide interchannel spacing was chosen to take advantage of the noise-suppressing feature of wideband FM.

Bandwidth of 200 kHz is not needed to accommodate an audio signal — 20 kHz to 30 kHz is all that is necessary for a narrowband FM signal. The 200 kHz bandwidth allowed room for ±75 kHz signal deviation from the assigned frequency, plus guard bands to reduce or eliminate adjacent channel interference. The larger bandwidth allows for broadcasting a 15 kHz bandwidth audio signal plus a 38 kHz stereo "subcarrier"—a piggyback signal that rides on the main signal. Additional unused capacity is used by some broadcasters to transmit utility functions such as background music for public areas, GPS auxiliary signals, or financial market data.

The AM radio problem of interference at night was addressed in a different way. At the time FM was set up, the available frequencies were far higher in the spectrum than those used for AM radio - by a factor of approximately 100. Using these frequencies meant that even at far higher power, the range of a given FM signal was much shorter; thus its market was more local than for AM radio. The reception range at night is the same as in the daytime. All FM broadcast transmissions are line-of-sight, and ionospheric bounce is not viable. The much larger bandwidths, compared to AM and SSB, are more susceptible to phase dispersion. Propagation speeds (celerities) are fastest in the ionosphere at the lowest sideband frequency. The celerity difference between the highest and lowest sidebands is quite apparent to the listener. Such distortion occurs up to frequencies of approximately 50 MHz. Higher frequencies do not reflect from the ionosphere, nor from storm clouds. Moon reflections have been used in some experiments, but require impractical power levels.

The original FM radio service in the U.S. was the Yankee Network, located in New England. Regular FM broadcasting began in 1939, but did not pose a significant threat to the AM broadcasting industry. It required purchase of a special receiver. The frequencies used, 42 to 50 MHz, were not those used today. The change to the current frequencies, 88 to 108 MHz, began after the end of World War II, and was to some extent imposed by AM broadcasters as an attempt to cripple what was by now realized to be a potentially serious threat.

FM radio on the new band had to begin from the ground floor. As a commercial venture it remained a little-used audio enthusiasts' medium until the 1960s. The more prosperous AM stations, or their owners, acquired FM licenses and often broadcast the same programming on the FM station as on the AM station ("simulcasting"). The FCC limited this practice in the 1960s. By the 1980s, since almost all new radios included both AM and FM tuners, FM became the dominant medium, especially in cities. Because of its greater range, AM remained more common in rural environments.

Pirate Radio

Pirate radio is illegal or non-regulated radio transmission. It is most commonly used to describe illegal broadcasting for entertainment or political purposes. Sometimes it is used for illegal two-way radio operation. Its history can be traced back to the unlicensed nature of the transmission, but historically there has been occasional use of sea vessels—fitting the most common perception of a pirate—as broadcasting bases. Rules and regulations vary largely from country to country, but often the term pirate radio generally describes the unlicensed broadcast of FM radio, AM radio, or short wave signals over a wide range. In some places radio stations are legal where the signal is transmitted, but illegal where the signals are received—especially when the signals cross a national boundary. In other cases, a broadcast may be considered "pirate" due to the type of content, its transmission format, or the transmitting power (wattage) of the station, even if the transmission is not technically illegal (such as a web cast or an amateur radio transmission). Pirate radio stations are sometimes referred to as bootleg radio or clandestine stations.

Terrestrial Digital Radio

Digital radio broadcasting has emerged, first in Europe (the UK in 1995 and Germany in 1999), and later in the United States, France, the Netherlands, South Africa and many other countries worldwide. The most simple system is named DAB Digital Radio, for Digital Audio Broadcasting, and uses the public domain EUREKA 147 (Band III) system. DAB is used mainly in the UK and

South Africa. Germany and the Netherlands use the DAB and DAB+ systems, and France uses the L-Band system of DAB Digital Radio.

In the United States, digital radio isn't used in the same way as Europe and South Africa. Instead, the IBOC system is named HD Radio and owned by a consortium of private companies that is called iBiquity. An international non-profit consortium Digital Radio Mondiale (DRM), has introduced the public domain DRM system.

Satellite

Satellite radio broadcasters are slowly emerging, but the enormous entry costs of space-based satellite transmitters, and restrictions on available radio spectrum licenses has restricted growth of this market. In the USA and Canada, just two services, XM Satellite Radio and Sirius Satellite Radio exist. Both XM and Sirius are owned by Sirius XM Radio, which was formed by the merger of XM and Sirius on July 29, 2008, whereas in Canada, XM Radio Canada and Sirius Canada remained separate companies until 2010. Worldspace in Africa and Asia, and MobaHO! in Japan and the ROK were two unsuccessful satellite radio operators which have gone out of business.

Program Formats

Radio program formats differ by country, regulation and markets. For instance, the U.S. Federal Communications Commission designates the 88–92 megahertz band in the U.S. for non-profit or educational programming, with advertising prohibited.

In addition, formats change in popularity as time passes and technology improves. Early radio equipment only allowed program material to be broadcast in real time, known as *live* broadcasting. As technology for sound recording improved, an increasing proportion of broadcast programming used pre-recorded material. A current trend is the automation of radio stations. Some stations now operate without direct human intervention by using entirely pre-recorded material sequenced by computer control.

References

- Nebeker, Frederik (2009). Dawn of the Electronic Age: Electrical Technologies in the Shaping of the Modern World, 1914 to 1945. John Wiley & Sons. pp. 14–15. ISBN 0470409746.
- Clint Smith, Curt Gervelis (2003). Wireless Network Performance Handbook. McGraw-Hill Professional. ISBN 0-07-140655-7.
- Free radio: electronic civil disobedience by Lawrence C. Soley. Published by Westview Press, 1998. ISBN 0-8133-9064-8, ISBN 978-0-8133-9064-2.
- Rebel Radio: The Full Story of British Pirate Radio by John Hind, Stephen Mosco. Published by Pluto Press, 1985. ISBN 0-7453-0055-3, ISBN 978-0-7453-0055-9.
- H. P. Westman et al., (ed), Reference Data for Radio Engineers, Fifth Edition, 1968, Howard W. Sams and Co., no ISBN, Library of Congress Card No. 43-14665.
- Demetrius T Paris and F. Kenneth Hurd, Basic Electromagnetic Theory, McGraw Hill, New York 1969 ISBN 0-07-048470-8.
- Radiowave propagation, edited by M.Hall and L.Barclay, page 2, published by Peter Peregrinus Ltd., (1989), ISBN 0-86341-156-8

- Radiowave propagation, edited by M.Hall and L.Barclay, published by Peter Peregrinus Ltd., page 3, (1989), ISBN 0-86341-156-8
- Clinton B. DeSoto (1936). 200 meters & Down - The Story of Amateur Radio. W. Hartford, CT: The American Radio Relay League. pp. 132–146. ISBN 0-87259-001-1.
- Davies, Kenneth (1990). Ionospheric Radio. IEE Electromagnetic Waves Series #31. London, UK: Peter Peregrinus Ltd/The Institution of Electrical Engineers. pp. 184–186. ISBN 0-86341-186-X.
- George Jacobs and Theodore J. Cohen, Shortwave Propagation Handbook. Hicksville, New York: CQ Publishing (1982), pp. 130-135. ISBN 978-0-943016-00-9.
- Ruey J. Sung & Michael R. Lauer (2000). Fundamental approaches to the management of cardiac arrhythmias. Springer. p. 153. ISBN 978-0-7923-6559-4.
- Melvin A. Shiffman, Sid J. Mirrafati, Samuel M. Lam and Chelso G. Cueteaux (2007). Simplified Facial Rejuvenation. Springer. p. 157. ISBN 978-3-540-71096-7.

4

Understanding Radio Frequency Identification

To understand radio frequency identification, it is very important to understand real-time locating system, microchip implant and transponder timing. Real-time locating systems are used in tracking objects and people in real-time. This section elucidates the crucial aspects of radio frequency identification.

Radio-frequency Identification

Radio-frequency identification (*RFID*) uses electromagnetic fields to automatically identify and track tags attached to objects. The tags contain electronically stored information. Passive tags collect energy from a nearby RFID reader's interrogating radio waves. Active tags have a local power source such as a battery and may operate at hundreds of meters from the RFID reader. Unlike a barcode, the tag need not be within the line of sight of the reader, so it may be embedded in the tracked object. RFID is one method for Automatic Identification and Data Capture (AIDC).

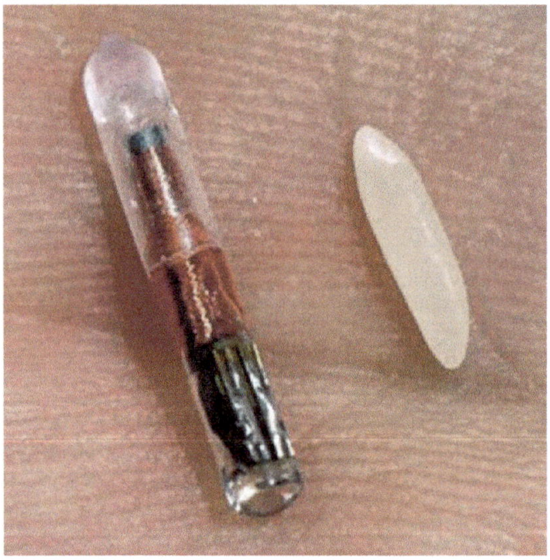

Small RFID chips, here compared to a grain of rice, are incorporated in consumer products, and implanted in pets, for identification purposes

RFID tags are used in many industries, for example, an RFID tag attached to an automobile during production can be used to track its progress through the assembly line; RFID-tagged pharmaceuticals can be tracked through warehouses; and implanting RFID microchips in livestock and pets allows positive identification of animals.

Since RFID tags can be attached to cash, clothing, and possessions, or implanted in animals and people, the possibility of reading personally-linked information without consent has raised serious privacy concerns. These concerns resulted in standard specifications development addressing privacy and security issues. ISO/IEC 18000 and ISO/IEC 29167 use on-chip cryptography methods for untraceability, tag and reader authentication, and over-the-air privacy. ISO/IEC 20248 specifies a digital signature data structure for RFID and barcodes providing data, source and read method authenticity. This work is done within ISO/IEC JTC 1/SC 31 Automatic identification and data capture techniques.

In 2014, the world RFID market is worth US$8.89 billion, up from US$7.77 billion in 2013 and US$6.96 billion in 2012. This includes tags, readers, and software/services for RFID cards, labels, fobs, and all other form factors. The market value is expected to rise to US$18.68 billion by 2026.

History

In 1945, Léon Theremin invented an espionage tool for the Soviet Union which retransmitted incident radio waves with the audio information. Sound waves vibrated a diaphragm which slightly altered the shape of the resonator, which modulated the reflected radio frequency. Even though this device was a covert listening device, not an identification tag, it is considered to be a predecessor of RFID, because it was likewise passive, being energized and activated by waves from an outside source.

Similar technology, such as the IFF transponder, was routinely used by the allies and Germany in World War II to identify aircraft as friend or foe. Transponders are still used by most powered aircraft to this day. Another early work exploring RFID is the landmark 1948 paper by Harry Stockman. Stockman predicted that "... considerable research and development work has to be done before the remaining basic problems in reflected-power communication are solved, and before the field of useful applications is explored."

Mario Cardullo's device, patented on January 23, 1973, was the first true ancestor of modern RFID, as it was a passive radio transponder with memory. The initial device was passive, powered by the interrogating signal, and was demonstrated in 1971 to the New York Port Authority and other potential users and consisted of a transponder with 16 bit memory for use as a toll device. The basic Cardullo patent covers the use of RF, sound and light as transmission media. The original business plan presented to investors in 1969 showed uses in transportation (automotive vehicle identification, automatic toll system, electronic license plate, electronic manifest, vehicle routing, vehicle performance monitoring), banking (electronic check book, electronic credit card), security (personnel identification, automatic gates, surveillance) and medical (identification, patient history).

An early demonstration of *reflected power* (modulated backscatter) RFID tags, both passive and semi-passive, was performed by Steven Depp, Alfred Koelle, and Robert Frayman at the Los Alamos National Laboratory in 1973. The portable system operated at 915 MHz and used 12-bit tags. This technique is used by the majority of today's UHFID and microwave RFID tags.

The first patent to be associated with the abbreviation RFID was granted to Charles Walton in 1983.

Design

Tags

A radio-frequency identification system uses *tags*, or *labels* attached to the objects to be identified. Two-way radio transmitter-receivers called *interrogators* or *readers* send a signal to the tag and read its response.

RFID tags can be either passive, active or battery-assisted passive. An active tag has an on-board battery and periodically transmits its ID signal. A battery-assisted passive (BAP) has a small battery on board and is activated when in the presence of an RFID reader. A passive tag is cheaper and smaller because it has no battery; instead, the tag uses the radio energy transmitted by the reader. However, to operate a passive tag, it must be illuminated with a power level roughly a thousand times stronger than for signal transmission. That makes a difference in interference and in exposure to radiation.

Tags may either be read-only, having a factory-assigned serial number that is used as a key into a database, or may be read/write, where object-specific data can be written into the tag by the system user. Field programmable tags may be write-once, read-multiple; "blank" tags may be written with an electronic product code by the user.

RFID tags contain at least two parts: an integrated circuit for storing and processing information, modulating and demodulating a radio-frequency (RF) signal, collecting DC power from the incident reader signal, and other specialized functions; and an antenna for receiving and transmitting the signal. The tag information is stored in a non-volatile memory. The RFID tag includes either fixed or programmable logic for processing the transmission and sensor data, respectively.

An RFID reader transmits an encoded radio signal to interrogate the tag. The RFID tag receives the message and then responds with its identification and other information. This may be only a unique tag serial number, or may be product-related information such as a stock number, lot or batch number, production date, or other specific information. Since tags have individual serial numbers, the RFID system design can discriminate among several tags that might be within the range of the RFID reader and read them simultaneously.

Readers

RFID systems can be classified by the type of tag and reader. A **Passive Reader Active Tag** (PRAT) system has a passive reader which only receives radio signals from active tags (battery operated, transmit only). The reception range of a PRAT system reader can be adjusted from 1–2,000 feet (0–600 m), allowing flexibility in applications such as asset protection and supervision.

An Active Reader Passive Tag (ARPT) system has an active reader, which transmits interrogator signals and also receives authentication replies from passive tags.

An Active Reader Active Tag (ARAT) system uses active tags awoken with an interrogator signal from the active reader. A variation of this system could also use a Battery-Assisted Passive (BAP) tag which acts like a passive tag but has a small battery to power the tag's return reporting signal.

Fixed readers are set up to create a specific interrogation zone which can be tightly controlled. This allows a highly defined reading area for when tags go in and out of the interrogation zone. Mobile readers may be hand-held or mounted on carts or vehicles.

Frequencies

RFID frequency bands						
Band	Regulations	Range	Data speed	ISO/IEC 18000 Section	Remarks	Approximate tag cost in volume (2006) US $
120–150 kHz (LF)	Unregulated	10 cm	Low	Part 2	Animal identification, factory data collection	$1
13.56 MHz (HF)	ISM band worldwide	10 cm–1 m	Low to moderate	Part 3	Smart cards (ISO/IEC 15693, ISO/IEC 14443 A,B). Non fully ISO compatible memory cards (Mifare Classic, iCLASS, Legic, Felica ...). Micro processor ISO compatible cards (Desfire EV1, Seos)	$0.50 to $5
433 MHz (UHF)	Short Range Devices	1–100 m	Moderate	Part 7	Defense applications, with active tags	$5
865-868 MHz (Europe) 902-928 MHz (North America) UHF	ISM band	1–12 m	Moderate to high	Part 6	EAN, various standards	$0.15 (passive tags)
2450-5800 MHz (microwave)	ISM band	1–2 m	High	Part 4	802.11 WLAN, Bluetooth standards	$25 (active tags)
3.1–10 GHz (microwave)	Ultra wide band	up to 200 m	High	Not Defined	requires semi-active or active tags	$5 projected

Signaling

Signaling between the reader and the tag is done in several different incompatible ways, depending on the frequency band used by the tag. Tags operating on LF and HF bands are, in terms of radio wavelength, very close to the reader antenna because they are only a small percentage of a wavelength away. In this near field region, the tag is closely coupled electrically with the transmitter in the reader. The tag can modulate the field produced by the reader by changing the electrical loading the tag represents. By switching between lower and higher relative loads, the tag produces a change that the reader can detect. At UHF and higher frequencies, the tag is more than one radio wavelength away from the reader, requiring a different approach. The tag can backscatter a signal. Active tags may contain functionally separated transmitters and receivers, and the tag need not respond on a frequency related to the reader's interrogation signal.

An Electronic Product Code (EPC) is one common type of data stored in a tag. When written into the tag by an RFID printer, the tag contains a 96-bit string of data. The first eight bits are a header which identifies the version of the protocol. The next 28 bits identify the organization that manages the data for this tag; the organization number is assigned by the EPCGlobal consortium. The next 24 bits are an object class, identifying the kind of product; the last 36 bits are a unique serial number for a particular tag. These last two fields are set by the organization that issued the tag. Rather like a URL, the total electronic product code number can be used as a key into a global database to uniquely identify a particular product.

Often more than one tag will respond to a tag reader, for example, many individual products with tags may be shipped in a common box or on a common pallet. Collision detection is important to allow reading of data. Two different types of protocols are used to "singulate" a particular tag, allowing its data to be read in the midst of many similar tags. In a slotted Aloha system, the reader broadcasts an initialization command and a parameter that the tags individually use to pseudo-randomly delay their responses. When using an "adaptive binary tree" protocol, the reader sends an initialization symbol and then transmits one bit of ID data at a time; only tags with matching bits respond, and eventually only one tag matches the complete ID string.

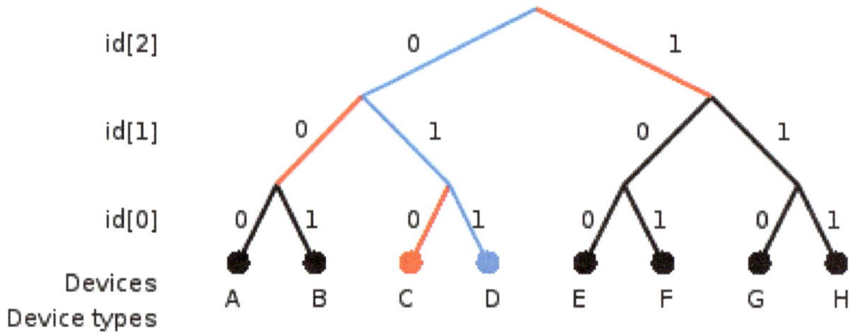

An example of a binary tree method of identifying an RFID tag

Both methods have drawbacks when used with many tags or with multiple overlapping readers. Bulk reading is a strategy for interrogating multiple tags at the same time, but lacks sufficient precision for inventory control.

Miniaturization

RFIDs are easy to conceal or incorporate in other items. For example, in 2009 researchers at Bristol University successfully glued RFID micro-transponders to live ants in order to study their behavior. This trend towards increasingly miniaturized RFIDs is likely to continue as technology advances.

Hitachi holds the record for the smallest RFID chip, at 0.05 mm × 0.05 mm. This is 1/64th the size of the previous record holder, the mu-chip. Manufacture is enabled by using the silicon-on-insulator (SOI) process. These dust-sized chips can store 38-digit numbers using 128-bit Read Only Memory (ROM). A major challenge is the attachment of antennas, thus limiting read range to only millimeters.

Uses

The RFID tag can be affixed to an object and used to track and manage inventory, assets, people, etc. For example, it can be affixed to cars, computer equipment, books, mobile phones, etc.

RFID offers advantages over manual systems or use of bar codes. The tag can be read if passed near a reader, even if it is covered by the object or not visible. The tag can be read inside a case, carton, box or other container, and unlike barcodes, RFID tags can be read hundreds at a time. Bar codes can only be read one at a time using current devices.

In 2011, the cost of passive tags started at US$0.09 each; special tags, meant to be mounted on metal or withstand gamma sterilization, can go up to US$5. Active tags for tracking containers, medical assets, or monitoring environmental conditions in data centers start at US$50 and can go up over US$100 each. Battery-Assisted Passive (BAP) tags are in the US$3–10 range and also have sensor capability like temperature and humidity.

RFID can be used in a variety of applications, such as:

Electronic Lock with RFID Card System, ANSI

Electronic key for RFID based lock system

- Access management
- Tracking of goods
- Tracking of persons and animals
- Toll collection and contactless payment

- Machine readable travel documents
- Smartdust (for massively distributed sensor networks)
- Tracking sports memorabilia to verify authenticity
- Airport baggage tracking logistics
- Timing sporting events

In 2010 three factors drove a significant increase in RFID usage: decreased cost of equipment and tags, increased performance to a reliability of 99.9% and a stable international standard around UHF passive RFID. The adoption of these standards were driven by EPCglobal, a joint venture between GS1 and GS1 US, which were responsible for driving global adoption of the barcode in the 1970s and 1980s. The EPCglobal Network was developed by the Auto-ID Center.

Commerce

An EPC RFID tag used by Wal-Mart.

Sewn-in RFID label in garment manufactured by the French sports supplier Decathlon. Front, back, and transparency scan.

RFID provides a way for organizations to identify and manage tools and equipment (asset tracking) , without manual data entry. RFID is being adopted for item level tagging in retail stores. This provides electronic article surveillance (EAS), and a self checkout process for consumers. Automatic identification with RFID can be used for inventory systems. Manufactured products such as automobiles or garments can be tracked through the factory and through shipping to the customer.

Casinos can use RFID to authenticate poker chips, and can selectively invalidate any chips known to be stolen.

Wal-Mart and the United States Department of Defense have published requirements that their vendors place RFID tags on all shipments to improve supply chain management.

Access Control

RFID antenna for vehicular access control

RFID tags are widely used in identification badges, replacing earlier magnetic stripe cards. These badges need only be held within a certain distance of the reader to authenticate the holder. Tags can also be placed on vehicles, which can be read at a distance, to allow entrance to controlled areas without having to stop the vehicle and present a card or enter an access code.

Advertising

In 2010 Vail Resorts began using UHF Passive RFID tags in ski passes. Facebook is using RFID cards at most of their live events to allow guests to automatically capture and post photos. The automotive brands have adopted RFID for social media product placement more quickly than other industries. Mercedes was an early adopter in 2011 at the PGA Golf Championships, and by the 2013 Geneva Motor Show many of the larger brands were using RFID for social media marketing.

Promotion Tracking

To prevent retailers diverting products, manufacturers are exploring the use of RFID tags on promoted merchandise so that they can track exactly which product has sold through the supply chain at fully discounted prices.

Transportation and Logistics

Yard management, shipping and freight and distribution centers use RFID tracking. In the railroad industry, RFID tags mounted on locomotives and rolling stock identify the owner, identification number and type of equipment and its characteristics. This can be used with a database to identify the lading, origin, destination, etc. of the commodities being carried.

In commercial aviation, RFID is used to support maintenance on commercial aircraft. RFID tags are used to identify baggage and cargo at several airports and airlines.

Some countries are using RFID for vehicle registration and enforcement. RFID can help detect and retrieve stolen cars.

Intelligent Transportation Systems

RFID E-ZPass reader attached to the pole and antenna (right) used in traffic monitoring in New York City

RFID is used in intelligent transportation systems. In New York City, RFID readers are deployed at intersections to track E-ZPass tags as a means for monitoring the traffic flow. The data is fed through the broadband wireless infrastructure to the traffic management center to be used in adaptive traffic control of the traffic lights.

Hose stations and Conveyance of Fluids

The RFID antenna in a permanently installed coupling half (fixed part) unmistakably identifies the RFID transponder placed in the other coupling half (free part) after completed coupling. When connected the transponder of the free part transmits all important information contactless to the fixed part. The coupling's location can be clearly identified by the RFID transponder coding. The control is enabled to automatically start subsequent process steps.

Track & Trace Test Vehicles and Prototype Parts

In the automotive industry RFID is used to Track & Trace test vehicles and prototype parts (project Transparent Prototype).

Public Transport

RFID cards are used for access control to public transport.

In London travellers use Oyster Cards on the tube, buses and ferries. It identifies the traveller at each turnstile and so the system can calculate the fare.

In the Chicago area, riders use the open standard Ventra card to board CTA buses and trains, along with PACE buses.

In Ontario, Canada, riders in the GTA and Ottawa Area use the Presto card to board trains, buses and street cars across multiple different transit companies.

Infrastructure Management and Protection

At least one company has introduced RFID to identify and locate underground infrastructure assets such as gas pipelines, sewer lines, electrical cables, communication cables, etc.

Passports

The first RFID passports ("E-passport") were issued by Malaysia in 1998. In addition to information also contained on the visual data page of the passport, Malaysian e-passports record the travel history (time, date, and place) of entries and exits from the country.

Other countries that insert RFID in passports include Norway (2005), Japan (March 1, 2006), most EU countries (around 2006), Australia, Hong Kong, the United States (2007), India (June 2008), Serbia (July 2008), Republic of Korea (August 2008), Taiwan (December 2008), Albania (January 2009), The Philippines (August 2009), Republic of Macedonia (2010), and Canada (2013).

Standards for RFID passports are determined by the International Civil Aviation Organization (ICAO), and are contained in ICAO Document 9303, Part 1, Volumes 1 and 2 (6th edition, 2006). ICAO refers to the ISO/IEC 14443 RFID chips in e-passports as "contactless integrated circuits". ICAO standards provide for e-passports to be identifiable by a standard e-passport logo on the front cover.

Since 2006, RFID tags included in new United States passports will store the same information that is printed within the passport, and include a digital picture of the owner. The United States Department of State initially stated the chips could only be read from a distance of 10 centimetres (3.9 in), but after widespread criticism and a clear demonstration that special equipment can read the test passports from 10 metres (33 ft) away, the passports were designed to incorporate a thin metal lining to make it more difficult for unauthorized readers to "skim" information when the passport is closed. The department will also implement Basic Access Control (BAC), which functions as a Personal Identification Number (PIN) in the form of characters printed on the passport data page. Before a passport's tag can be read, this PIN must be entered into an RFID reader. The BAC also enables the encryption of any communication between the chip and interrogator. As noted in the section below on security, there are many situations in which these protections have been shown to be insufficient, and passports have been cloned based on scans of them while they were being delivered in the mail.

Transportation Payments

In many countries, RFID tags can be used to pay for mass transit fares on bus, trains, or subways, or to collect tolls on highways.

Some bike lockers are operated with RFID cards assigned to individual users. A prepaid card is required to open or enter a facility or locker and is used to track and charge based on how long the bike is parked.

The Zipcar car-sharing service uses RFID cards for locking and unlocking cars and for member identification.

In Singapore, RFID replaces paper Season Parking Ticket (SPT).

Animal Identification

RFID tags for animals represent one of the oldest uses of RFID. Originally meant for large ranches and rough terrain, since the outbreak of mad-cow disease, RFID has become crucial in animal identification management. An implantable RFID tag or transponder can also be used for animal identification. The transponders are better known as PIT (Passive Integrated Transponder) tags, passive RFID, or "chips" on animals. The Canadian Cattle Identification Agency began using RFID tags as a replacement for barcode tags. Currently CCIA tags are used in Wisconsin and by United States farmers on a voluntary basis. The USDA is currently developing its own program.

RFID tags are required for all cattle, sheep and goats sold in Australia.

Human Identification

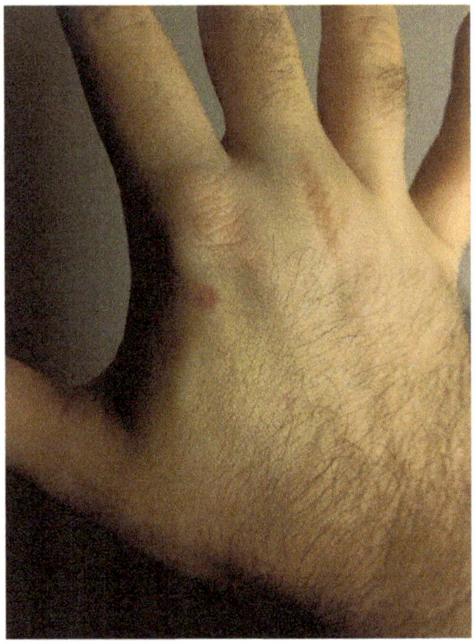

An RFID microchip implant

Implantable RFID chips designed for animal tagging are now being used in humans. An early experiment with RFID implants was conducted by British professor of cybernetics Kevin Warwick, who implanted a chip in his arm in 1998. In 2004 Conrad Chase offered implanted chips in his night clubs in Barcelona and Rotterdam to identify their VIP customers, who in turn use it to pay for drinks.

The Food and Drug Administration in the United States has approved the use of RFID chips in humans. Some business establishments give customers the option of using an RFID-based tab to pay for service, such as the *Baja Beach* nightclub in Barcelona. This has provoked concerns into privacy of individuals as they can potentially be tracked wherever they go by an identifier unique to them. Some are concerned this could lead to abuse by an authoritarian government, to removal of freedoms, and to the emergence of an "ultimate panopticon", a society where all citizens behave in a socially accepted manner because others might be watching.

On July 22, 2006, Reuters reported that two hackers, Newitz and Westhues, at a conference in New York City showed that they could clone the RFID signal from a human implanted RFID chip, showing that the chip is not hack-proof as was previously claimed. Privacy advocates have protested against implantable RFID chips, warning of potential abuse. There is much controversy regarding human applications of this technology, and many conspiracy theories abound in relation to human applications, especially one of which is referred to as "The Mark of the Beast" in some religious circles.

Institutions

Hospitals and Healthcare

In Healthcare, there is a need for increased visibility, efficiency, and gathering of data around relevant interactions. RFID tracking solutions are able to help healthcare facilities manage mobile medical equipment, improve patient workflow, monitor environmental conditions, and protect patients, staff and visitors from infection or other hazards.

Adoption of RFID in the medical industry has been widespread and very effective. Hospitals are among the first users to combine both active and passive RFID. Many successful deployments in the healthcare industry have been cited where active technology tracks high-value, or frequently moved items, where passive technology tracks smaller, lower cost items that only need room-level identification. For example, medical facility rooms can collect data from transmissions of RFID badges worn by patients and employees, as well as from tags assigned to facility assets, such as mobile medical devices. The U.S. Department of Veterans Affairs (VA) recently announced plans to deploy RFID in hospitals across America to improve care and reduce costs.

A physical RFID tag may be incorporated with browser-based software to increase its efficacy. This software allows for different groups or specific hospital staff, nurses, and patients to see real-time data relevant to each piece of tracked equipment or personnel. Real-time data is stored and archived to make use of historical reporting functionality and to prove compliance with various industry regulations. This combination of RFID real-time locating system hardware and software provides a powerful data collection tool for facilities seeking to improve operational efficiency and reduce costs.

The trend is toward using ISO 18000-6c as the tag of choice and combining an active tagging system that relies on existing 802.11X wireless infrastructure for active tags.

Since 2004 a number of U.S. hospitals have begun implanting patients with RFID tags and using RFID systems, usually for workflow and inventory management. The use of RFID to prevent mix-ups between sperm and ova in IVF clinics is also being considered.

In October 2004, the FDA approved USA's first RFID chips that can be implanted in humans. The 134 kHz RFID chips, from VeriChip Corp. can incorporate personal medical information and could save lives and limit injuries from errors in medical treatments, according to the company. Anti-RFID activists Katherine Albrecht and Liz McIntyre discovered an FDA Warning Letter that spelled out health risks. According to the FDA, these include "adverse tissue reaction", "migration of the implanted transponder", "failure of implanted transponder", "electrical hazards" and "magnetic resonance imaging [MRI] incompatibility."

Libraries

RFID tags used in libraries: square book tag, round CD/DVD tag and rectangular VHS tag

Libraries have used RFID to replace the barcodes on library items. The tag can contain identifying information or may just be a key into a database. An RFID system may replace or supplement bar codes and may offer another method of inventory management and self-service checkout by patrons. It can also act as a security device, taking the place of the more traditional electromagnetic security strip.

It is estimated that over 30 million library items worldwide now contain RFID tags, including some in the Vatican Library in Rome.

Since RFID tags can be read through an item, there is no need to open a book cover or DVD case to scan an item, and a stack of books can be read simultaneously. Book tags can be read while books are in motion on a conveyor belt, which reduces staff time. This can all be done by the borrowers themselves, reducing the need for library staff assistance. With portable readers, inventories could be done on a whole shelf of materials within seconds. However, as of 2008 this technology remains too costly for many smaller libraries, and the conversion period has been estimated at 11 months for an average-size library. A 2004 Dutch estimate was that a library which lends 100,000 books per year should plan on a cost of €50,000 (borrow- and return-stations: 12,500 each, detection porches 10,000 each; tags 0.36 each). RFID taking a large burden off staff could also mean that fewer staff will be needed, resulting in some of them getting laid off, but that has so far not happened in North America where recent surveys have not returned a single library that cut staff because of adding RFID. In fact, library budgets are being reduced for personnel and increased for infrastructure, making it necessary for libraries to add automation to compensate for the reduced staff size. Also, the tasks that RFID takes over are largely not the primary tasks of librarians. A finding in the Netherlands is that borrowers are pleased with the fact that staff are now more available for answering questions.

Privacy concerns have been raised surrounding library use of RFID. Because some RFID tags can be read from up to 100 metres (330 ft), there is some concern over whether sensitive information could be collected from an unwilling source. However, library RFID tags do not contain any patron information, and the tags used in the majority of libraries use a frequency only readable from approximately 10 feet (3.0 m). Further, another non-library agency could potentially record the RFID tags of every person leaving the library without the library administrator's knowledge or consent. One simple option is to let the book transmit a code that has meaning only in conjunction with the library's database. Another possible enhancement would be to give each book a new code every time it is returned. In future, should readers become ubiquitous (and possibly networked), then stolen books could be traced even outside the library. Tag removal could be made difficult if the tags are so small that they fit invisibly inside a (random) page, possibly put there by the publisher.

Museums

RFID technologies are now also implemented in end-user applications in museums. An example was the custom-designed temporary research application, "eXspot," at the Exploratorium, a science museum in San Francisco, California. A visitor entering the museum received an RF Tag that could be carried as a card. The eXspot system enabled the visitor to receive information about specific exhibits. Aside from the exhibit information, the visitor could take photographs of themselves at the exhibit. It was also intended to allow the visitor to take data for later analysis. The collected information could be retrieved at home from a "personalized" website keyed to the RFID tag.

Schools and Universities

School authorities in the Japanese city of Osaka are now chipping children's clothing, backpacks, and student IDs in a primary school. A school in Doncaster, England is piloting a monitoring system designed to keep tabs on pupils by tracking radio chips in their uniforms. St Charles Sixth Form College in west London, England, started September, 2008, is using an RFID card system to check in and out of the main gate, to both track attendance and prevent unauthorized entrance. Similarly, Whitcliffe Mount School in Cleckheaton, England uses RFID to track pupils and staff in and out of the building via a specially designed card. In the Philippines, some schools already use RFID in IDs for borrowing books and also gates in those particular schools have RFID ID scanners for buying items at a school shop and canteen, library and also to sign in and sign out for student and teacher's attendance.

Sports

RFID for timing races began in the early 1990s with pigeon racing, introduced by the company Deister Electronics in Germany. RFID can provide race start and end timings for individuals in large races where it is impossible to get accurate stopwatch readings for every entrant.

ChampionChip

In the race, the racers wear tags that are read by antennas placed alongside the track or on mats across the track. UHF tags provide accurate readings with specially designed antennas. Rush error, lap count errors and accidents at start time are avoided since anyone can start and finish any time without being in a batch mode.

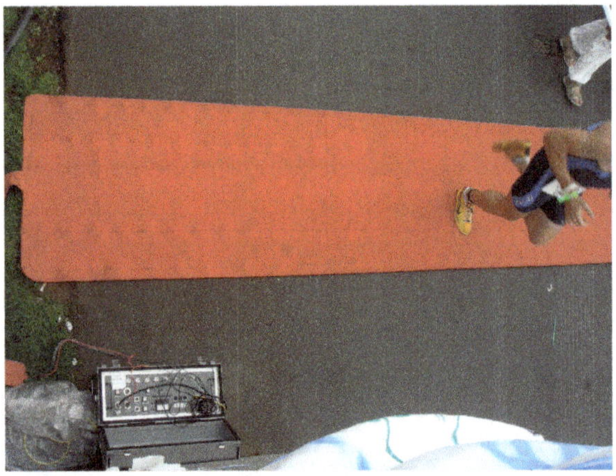

J-Chip 8-channel receiver next to timing mat. The athlete wears a chip on a strap around his ankle. Ironman Germany 2007 in Frankfurt.

The design of chip+antenna controls the range from which it can be read. Short range compact chips are twist tied to the shoe or velcro strapped the ankle. These need to be about 400mm from the mat and so give very good temporal resolution. Alternatively a chip plus a very large (a 125mm square) antenna can be incorporated into the bib number worn on the athlete's chest at about 1.25m height.

Passive and active RFID systems are used in off-road events such as Orienteering, Enduro and Hare and Hounds racing. Riders have a transponder on their person, normally on their arm. When they complete a lap they swipe or touch the receiver which is connected to a computer and log their lap time.

RFID is being adapted by many recruitment agencies which have a PET (Physical Endurance Test) as their qualifying procedure especially in cases where the candidate volumes may run into millions (Indian Railway Recruitment Cells, Police and Power sector).

A number of ski resorts have adopted RFID tags to provide skiers hands-free access to ski lifts. Skiers do not have to take their passes out of their pockets. Ski jackets have a left pocket into which the chip+card fits. This nearly contacts the sensor unit on the left of the turnstile as the skier pushes through to the lift. These systems were based on high frequency (HF) at 13.56 megahertz. The bulk of ski areas in Europe, from Verbier to Chamonix use these systems.

The NFL in the United States equips players with RFID chips that measures speed, distance and direction traveled by each player in real-time. Currently cameras stay focused on the quarterback, however, numerous plays are happening simultaneously on the field. The RFID chip will provide new insight into these simultaneous plays. The chip triangulates the player's position within six inches and will be used to digitally broadcast replays. The RFID chip will make individual player information accessible to the public. The data will be available via the NFL 2015 app. The RFID chips are manufactured by Zebra Technologies. Zebra Technologies tested the RFID chip in 18 stadiums last year to track vector data.

Complement to Barcode

RFID tags are often a complement, but not a substitute, for UPC or EAN barcodes. They may never completely replace barcodes, due in part to their higher cost and the advantage of multiple data sources on the same object. Also, unlike RFID labels, barcodes can be generated and distributed electronically, *e.g.* via e-mail or mobile phone, for printing or display by the recipient. An example is airline boarding passes. The new EPC, along with several other schemes, is widely available at reasonable cost.

The storage of data associated with tracking items will require many terabytes. Filtering and categorizing RFID data is needed to create useful information. It is likely that goods will be tracked by the pallet using RFID tags, and at package level with Universal Product Code (UPC) or EAN from unique barcodes.

The unique identity is a mandatory requirement for RFID tags, despite special choice of the numbering scheme. RFID tag data capacity is large enough that each individual tag will have a unique code, while current bar codes are limited to a single type code for a particular product. The uniqueness of RFID tags means that a product may be tracked as it moves from location to location, finally ending up in the consumer's hands. This may help to combat theft and other forms of product loss. The tracing of products is an important feature that gets well supported with RFID tags containing a unique identity of the tag and also the serial number of the object. This may help companies cope with quality deficiencies and resulting recall campaigns, but also contributes to concern about tracking and profiling of consumers after the sale.

Telemetry

Active RFID tags also have the potential to function as low-cost remote sensors that broadcast telemetry back to a base station. Applications of tagometry data could include sensing of road conditions by implanted beacons, weather reports, and noise level monitoring.

Passive RFID tags can also report sensor data. For example, the Wireless Identification and Sensing Platform is a passive tag that reports temperature, acceleration and capacitance to commercial Gen2 RFID readers.

It is possible that active or battery-assisted passive (BAP) RFID tags, could broadcast a signal to an in-store receiver to determine whether the RFID tag (product) is in the store.

Optical RFID

Optical RFID (aka OPID) is an alternative to RFID that is based on optical readers. Applications for optical RFID tags may be found in future supply chain scenarios. The main advantage in comparison to traditional RFID tags is their low price and the usually employed offline preaggregation of data to the class level.

Unlike most other RFID chips (which use radio frequencies of 0.125–0.1342, 0.140–0.1485, 13.56, and 868–928 MHz), optical RFID operates in the electromagnetic spectrum between the frequencies of 333 THz (3.33×10^{14} hertz, 900 nm) and 380 THz (788 nm) and 750 THz (400 nm). The tag information is communicated to the reader by reflecting the read request. Parts of the incoming signal are filtered by the tag in a well-defined way as it is sent back to the reader. On the reader's side, the tag data can be deduced by analysing the pattern used for filtering. As an alternative to reflection mode, active circuits can be used, replacing awkward RFID antennae with photovoltaic components and IR-LEDs on the ICs.One of the earliest examples of Optical RFID is RFIG: Radio Frequency Identity and Geometry, by Ramesh Raskar, Paul Dietz, Paul Beardsley and colleagues. This combines an optical tag with a RF tag to provide ID as well as geometric operations, such as location, pose, motion, orientation and change detection.

Regarding privacy, optical RFID provides much more protection against abuse than RFID based on common electromagnetic waves. This is mainly because line-of-sight is required for malicious read out. Such an attack can easily be prevented with low cost optical RFID sight blockers. Nevertheless, if needed some penetration of solids and liquids can be achieved e.g. with near-IR wavelengths.

Regulation and Standardization

A number of organizations have set standards for RFID, including the International Organization for Standardization (ISO), the International Electrotechnical Commission (IEC), ASTM International, the DASH7 Alliance and EPCglobal.

There are also several specific industries that have set guidelines. These industries include the Financial Services Technology Consortium (FSTC) which has set a standard for tracking IT Assets with RFID, the Computer Technology Industry Association CompTIA which has set a standard for certifying RFID engineers, and the International Airlines Transport Association IATA which has set tagging guidelines for luggage in airports.

In principle, every country can set its own rules for frequency allocation for RFID tags, and not all radio bands are available in all countries. These frequencies are known as the ISM bands (Industrial Scientific and Medical bands). The return signal of the tag may still cause interference for other radio users.

- Low-frequency (LF: 125–134.2 kHz and 140–148.5 kHz) (LowFID) tags and high-frequency (HF: 13.56 MHz) (HighFID) tags can be used globally without a license.

- Ultra-high-frequency (UHF: 865–928 MHz) (Ultra-HighFID or UHFID) tags cannot be used globally as there is no single global standard and regulations differ from country to country.

In North America, UHF can be used unlicensed for 902–928 MHz (±13 MHz from the 915 MHz center frequency), but restrictions exist for transmission power. In Europe, RFID and other low-power radio applications are regulated by ETSI recommendations EN 300 220 and EN 302 208, and ERO recommendation 70 03, allowing RFID operation with somewhat complex band restrictions from 865–868 MHz. Readers are required to monitor a channel before transmitting ("Listen Before Talk"); this requirement has led to some restrictions on performance, the resolution of which is a subject of current research. The North American UHF standard is not accepted in France as it interferes with its military bands. On July 25, 2012, Japan changed its UHF band to 920 MHz, more closely matching the United States' 915 MHz band.

In some countries, a site license is needed, which needs to be applied for at the local authorities, and can be revoked.

According to an overview assembled by GS1, as of 31 October 2014, regulations are in place in 78 countries representing ca. 96.5% of the world's GDP, and work on regulations is in progress in 3 countries representing circa 1% of the world's GDP.

Standards that have been made regarding RFID include:

- ISO 14223 – Radiofrequency [*sic*] identification of animals – Advanced transponders
- ISO/IEC 14443: This standard is a popular HF (13.56 MHz) standard for HighFIDs which is being used as the basis of RFID-enabled passports under ICAO 9303. The Near Field Communication standard that lets mobile devices act as RFID readers/transponders is also based on ISO/IEC 14443.
- ISO/IEC 15693: This is also a popular HF (13.56 MHz) standard for HighFIDs widely used for non-contact smart payment and credit cards.
- ISO/IEC 18000: Information technology—Radio frequency identification for item management:
    - Part 1: Reference architecture and definition of parameters to be standardized
    - Part 2: Parameters for air interface communications below 135 kHz
    - Part 3: Parameters for air interface communications at 13.56 MHz
    - Part 4: Parameters for air interface communications at 2.45 GHz
    - Part 6: Parameters for air interface communications at 860–960 MHz
    - Part 7: Parameters for active air interface communications at 433 MHz
- ISO/IEC 18092 Information technology—Telecommunications and information exchange between systems—Near Field Communication—Interface and Protocol (NFCIP-1)
- ISO 18185: This is the industry standard for electronic seals or "e-seals" for tracking cargo containers using the 433 MHz and 2.4 GHz frequencies.

- ISO/IEC 21481 Information technology—Telecommunications and information exchange between systems—Near Field Communication Interface and Protocol -2 (NFCIP-2)
- ASTM D7434, Standard Test Method for Determining the Performance of Passive Radio Frequency Identification (RFID) Transponders on Palletized or Unitized Loads
- ASTM D7435, Standard Test Method for Determining the Performance of Passive Radio Frequency Identification (RFID) Transponders on Loaded Containers
- ASTM D7580, Standard Test Method for Rotary Stretch Wrapper Method for Determining the Readability of Passive RFID Transponders on Homogenous Palletized or Unitized Loads
- ISO 28560-2 : specifies encoding standards and data model to be used within libraries.

In order to ensure global interoperability of products, several organizations have set up additional standards for RFID testing. These standards include conformance, performance and interoperability tests.

Groups concerned with standardization are:

- DASH7 Alliance – an international industry group formed in 2009 to promote standards and interoperability among extensions to ISO/IEC 18000-7 technologies
- EPCglobal – this is the standardization framework that is most likely to undergo international standardisation according to ISO rules as with all sound standards in the world, unless residing with limited scope, as customs regulations, air-traffic regulations and others. Currently the big distributors and governmental customers are pushing EPC heavily as a standard well-accepted in their community, but not yet regarded as for salvation to the rest of the world.

EPC Gen2

EPC Gen2 is short for *EPCglobal UHF Class 1 Generation 2*.

EPCglobal, a joint venture between GS1 and GS1 US, is working on international standards for the use of mostly passive RFID and the Electronic Product Code (EPC) in the identification of many items in the supply chain for companies worldwide.

One of the missions of EPCglobal was to simplify the Babel of protocols prevalent in the RFID world in the 1990s. Two tag air interfaces (the protocol for exchanging information between a tag and a reader) were defined (but not ratified) by EPCglobal prior to 2003. These protocols, commonly known as Class 0 and Class 1, saw significant[*clarification needed*] commercial implementation in 2002–2005.

In 2004, the Hardware Action Group created a new protocol, the Class 1 Generation 2 interface, which addressed a number of problems that had been experienced with Class 0 and Class 1 tags. The EPC Gen2 standard was approved in December 2004. This was approved after a contention from Intermec that the standard may infringe a number of their RFID-related patents. It was decided that the standard itself does not infringe their patents, making the standard royalty free. The EPC Gen2 standard was adopted with minor modifications as ISO 18000-6C in 2006.

In 2007, the lowest cost of Gen2 EPC inlay was offered by the now-defunct company SmartCode, at a price of $0.05 apiece in volumes of 100 million or more. Nevertheless, further conversion (including additional label stock or encapsulation processing/insertion and freight costs to a given facility or DC) and of the inlays into usable RFID labels and the design of current Gen 2 protocol standard will increase the total end-cost, especially with the added security feature extensions for RFID Supply Chain item-level tagging.

Problems and Concerns

Data Flooding

Not every successful reading of a tag (an observation) is useful for business purposes. A large amount of data may be generated that is not useful for managing inventory or other applications. For example, a customer moving a product from one shelf to another, or a pallet load of articles that passes several readers while being moved in a warehouse, are events that do not produce data that is meaningful to an inventory control system.

Event filtering is required to reduce this data inflow to a meaningful depiction of moving goods passing a threshold. Various concepts have been designed, mainly offered as *middleware* performing the filtering from noisy and redundant raw data to significant processed data.

Global Standardization

The frequencies used for UHF RFID in the USA are currently incompatible with those of Europe or Japan. Furthermore, no emerging standard has yet become as universal as the barcode. To address international trade concerns, it is necessary to use a tag that is operational within all of the international frequency domains.

Security Concerns

Retailers such as Walmart, which already heavily use RFID for inventory purposes, also use RFID as an anti-employee-theft and anti-shoplifting technology. If a product with an active RFID tag passes the exit-scanners at a Walmart outlet, not only does it set off an alarm, but it also tells security personnel exactly what product to look for in the shopper's cart.

A primary RFID security concern is the illicit tracking of RFID tags. Tags, which are world-readable, pose a risk to both personal location privacy and corporate/military security. Such concerns have been raised with respect to the United States Department of Defense's recent adoption of RFID tags for supply chain management. More generally, privacy organizations have expressed concerns in the context of ongoing efforts to embed electronic product code (EPC) RFID tags in consumer products. This is mostly as result of the fact that RFID tags can be read, and legitimate transactions with readers can be eavesdropped, from non-trivial distances. RFID used in access control, payment and eID (e-passport) systems operate at a shorter range than EPC RFID systems but are also vulnerable to skimming and eavesdropping, albeit at shorter distance.

A second method of prevention is by using cryptography. Rolling codes and challenge-response authentication (CRA) are commonly used to foil monitor-repetition of the messages between the tag and reader; as any messages that have been recorded would prove to be unsuccessful on repeat transmission. Rolling codes rely upon the tag's id being changed after each interrogation, while CRA uses software to ask for a cryptographically coded response from the tag. The protocols used during CRA can be symmetric, or may use public key cryptography.

Security concerns exist in regard to privacy over the unauthorized reading of RFID tags. Unauthorized readers can potentially use RFID information to identify or track packages, consumers, carriers, or the contents of a package. Several prototype systems are being developed to combat unauthorized reading, including RFID signal interruption, as well as the possibility of legislation, and 700 scientific papers have been published on this matter since 2002. There are also concerns that the database structure of Object Naming Services may be susceptible to infiltration, similar to denial-of-service attacks, after the EPCglobal Network ONS root servers were shown to be vulnerable.

Exploitation

Ars Technica reported in March 2006 an RFID buffer overflow bug that could infect airport terminal RFID databases for baggage, and also passport databases to obtain confidential information on the passport holder.

Passports

In an effort to standardize and make it easier to process passports, several countries have implemented RFID in passports, despite security and privacy issues. The encryption on UK chips was broken in under 48 hours. Since that incident, further efforts have allowed researchers to clone passport data while the passport is being mailed to its owner. Where a criminal used to need to secretly open and then reseal the envelope, now it can be done without detection, adding some degree of insecurity to the passport system.

Shielding

In an effort to prevent the passive "skimming" of RFID-enabled cards or passports, the U.S. General Services Administration (GSA) issued a set of test procedures for evaluating electromagnetically opaque sleeves. For shielding products to be in compliance with FIPS-201 guidelines, they must meet or exceed this published standard. Shielding products currently evaluated as FIPS-201 compliant are listed on the website of the U.S. CIO's FIPS-201 Evaluation Program. The United States government requires that when new ID cards are issued, they must be delivered with an approved shielding sleeve or holder.

There are contradicting opinions as to whether aluminum can prevent reading of RFID chips. Some people claim that aluminum shielding, essentially creating a Faraday cage, does work. Others claim that simply wrapping an RFID card in aluminum foil only makes transmission more difficult and is not completely effective at preventing it.

Shielding effectiveness depends on the frequency being used. Low-frequency LowFID tags, like those used in implantable devices for humans and pets, are relatively resistant to shielding though thick metal foil will prevent most reads. High frequency HighFID tags (13.56 MHz—smart cards and access badges) are sensitive to shielding and are difficult to read when within a few centimetres of a metal surface. UHF Ultra-HighFID tags (pallets and cartons) are difficult to read when placed within a few millimetres of a metal surface, although their read range is actually increased when they are spaced 2–4 cm from a metal surface due to positive reinforcement of the reflected wave and the incident wave at the tag.

Controversies

Logo of the anti-RFID campaign by German privacy group digitalcourage (formerly FoeBuD).

Privacy

The use of RFID has engendered considerable controversy and even product boycotts by consumer privacy advocates. Consumer privacy experts Katherine Albrecht and Liz McIntyre are two prominent critics of the "spychip" technology. The two main privacy concerns regarding RFID are:

- Since the owner of an item will not necessarily be aware of the presence of an RFID tag and the tag can be read at a distance without the knowledge of the individual, it becomes possible to gather sensitive data about an individual without consent.
- If a tagged item is paid for by credit card or in conjunction with use of a loyalty card, then it would be possible to indirectly deduce the identity of the purchaser by reading the globally unique ID of that item (contained in the RFID tag). This is only true if the person doing the watching also had access to the loyalty card data and the credit card data, and the person with the equipment knows where you are going to be.

Most concerns revolve around the fact that RFID tags affixed to products remain functional even after the products have been purchased and taken home and thus can be used for surveillance and other purposes unrelated to their supply chain inventory functions.

The RFID Network argued that these fears are unfounded in the first episode of their syndicated cable TV series by letting RF engineers demonstrate how RFID works. They provided images of RF engineers driving an RFID-enabled van around a building and trying to take an inventory of items inside. They discussed satellite tracking of a passive RFID tag, which is surprising since the maximum range is under 200m.

The concerns raised by the above may be addressed in part by use of the Clipped Tag. The Clipped Tag is an RFID tag designed to increase consumer privacy. The Clipped Tag has been suggested by IBM researchers Paul Moskowitz and Guenter Karjoth. After the point of sale, a consumer may tear off a portion of the tag. This allows the transformation of a long-range tag into a proximity tag that still may be read, but only at short range – less than a few inches or centimeters. The modification of the tag may be confirmed visually. The tag may still be used later for returns, recalls, or recycling.

However, read range is both a function of the reader and the tag itself. Improvements in technology may increase read ranges for tags. Tags may be read at longer ranges than they are designed for by increasing reader power. The limit on read distance then becomes the signal-to-noise ratio of the signal reflected from the tag back to the reader. Researchers at two security conferences have demonstrated that passive Ultra-HighFID tags normally read at ranges of up to 30 feet, can be read at ranges of 50 to 69 feet using suitable equipment.

In January 2004 privacy advocates from CASPIAN and the German privacy group FoeBuD were invited to the METRO Future Store in Germany, where an RFID pilot project was implemented. It was uncovered by accident that METRO "Payback" customer loyalty cards contained RFID tags with customer IDs, a fact that was disclosed neither to customers receiving the cards, nor to this group of privacy advocates. This happened despite assurances by METRO that no customer identification data was tracked and all RFID usage was clearly disclosed.

During the UN World Summit on the Information Society (WSIS) between the 16th to 18 November 2005, founder of the free software movement, Richard Stallman, protested the use of RFID security cards by covering his card with aluminum foil.

In 2004–2005 the Federal Trade Commission Staff conducted a workshop and review of RFID privacy concerns and issued a report recommending best practices.

RFID was one of the main topics of 2006 Chaos Communication Congress (organized by the Chaos Computer Club in Berlin) and triggered a big press debate. Topics included: electronic passports, Mifare cryptography and the tickets for the FIFA World Cup 2006. Talks showed how the first real world mass application of RFID at the 2006 FIFA Soccer World Cup worked. Group monochrom staged a special 'Hack RFID' song.

Government Control

Some individuals have grown to fear the loss of rights due to RFID human implantation.

By early 2007, Chris Paget of San Francisco, California, showed that RFID information can be pulled from individuals by using only $250 worth of equipment. This supports the claim that with the information captured, it would be relatively simple to make counterfeit passports.

According to ZDNet, critics believe that RFID will lead to tracking individuals' every movement and will be an invasion of privacy. In the book *SpyChips: How Major Corporations and Government Plan to Track Your Every Move* by Katherine Albrecht and Liz McIntyre, one is encouraged to "imagine a world of no privacy. Where your every purchase is monitored and recorded in a database and your every belonging is numbered. Where someone many states away or perhaps in another country has a record of everything you have ever bought. What's more, they can be tracked and monitored remotely".

Deliberate Destruction in Clothing and other Items

According to an RSA laboratories FAQ, RFID tags can be destroyed by a standard microwave oven; however some types of RFID tags, particularly those constructed to radiate using large metallic antennas (in particular RF tags and EPC tags), may catch fire if subjected to this process for too long (as would any metallic item inside a microwave oven). This simple method cannot safely be used to deactivate RFID features in electronic devices, or those implanted in living tissue, because of the risk of damage to the "host". However the time required is extremely short (a second or two of radiation) and the method works in many other non-electronic and inanimate items, long before heat or fire become of concern.

Some RFID tags implement a "kill command" mechanism for permanently and irreversibly disabling them. This mechanism can be applied if the chip itself is trusted or the mechanism is known by the person that wants to "kill" the tag.

UHF RFID tags that comply with the EPC2 Gen 2 Class 1 standard usually support this mechanism, while protecting the chip from being killed with a password. Guessing or cracking this needed 32-bit password for killing a tag would not be difficult for a determined attacker.

Real-time Locating System

Real-time locating systems (RTLS) are used to automatically identify and track the location of objects or people in real time, usually within a building or other contained area. Wireless RTLS tags are attached to objects or worn by people, and in most RTLS, fixed reference points receive wireless signals from tags to determine their location. Examples of real-time locating systems include tracking automobiles through an assembly line, locating pallets of merchandise in a warehouse, or finding medical equipment in a hospital.

The physical layer of RTLS technology is usually some form of radio frequency (RF) communication, but some systems use optical (usually infrared) or acoustic (usually ultrasound) technology instead of or in addition to RF. Tags and fixed reference points can be transmitters, receivers, or both, resulting in numerous possible technology combinations.

RTLS are a form of local positioning system, and do not usually refer to GPS or to mobile phone tracking. Location information usually does not include speed, direction, or spatial orientation.

Origin

The term RTLS was created (circa 1998) at the ID EXPO trade show by Tim Harrington (WhereNet), Jay Werb, (PinPoint), and Bert Moore, (Automatic Identification Manufacturers, Inc.(AIM)). It was created to describe and differentiate an emerging technology that not only provided the automatic identification capabilities of active RFID tags, but also added the ability to view the location on a computer screen. It was at this show that the first examples of a commercial radio based RTLS system were shown by PinPoint and WhereNet. Although this capability had been utilized previously by military and government agencies, the technology had been too expensive for commercial purposes. In the early 1990s, the first commercial RTLS were installed at three healthcare facilities in the United States, and were based on the transmission and decoding of infrared light signals from actively transmitting tags. Since then, new technology has emerged that also enables RTLS to be applied to passive tag applications.

Locating Concepts

RTLS are generally used in indoor and/or confined areas, such as buildings, and do not provide global coverage like GPS. RTLS tags are affixed to mobile items to be tracked or managed. RTLS reference points, which can be either transmitters or receivers, are spaced throughout a building (or similar area of interest) to provide the desired tag coverage. In most cases, the more RTLS reference points that are installed, the better the location accuracy, until the technology limitations are reached.

A number of disparate system designs are all referred to as "real-time locating systems", but there are two primary system design elements:

Locating at Choke Points

The simplest form of choke point locating is where short range ID signals from a moving tag are received by a single fixed reader in a sensory network, thus indicating the location coincidence of reader and tag. Alternately, a choke point identifier can be received by the moving tag, and then relayed, usually via a second wireless channel, to a location processor. Accuracy is usually defined by the sphere spanned with the reach of the choke point transmitter or receiver. The use of directional antennas, or technologies such as infrared or ultrasound that are blocked by room partitions, can support choke points of various geometries.

Locating in Relative Coordinates

ID signals from a tag is received by a multiplicity of readers in a sensory network, and a position is estimated using one or more locating algorithms, such as trilateration, multilateration, or triangulation. Equivalently, ID signals from several RTLS reference points can be received by a tag, and relayed back to a location processor. Localization with multiple reference points requires that distances between reference points in the sensory network be known in order to precisely locate a tag, and the determination of distances is called ranging.

Another way to calculate relative location is if mobile tags communicate directly with each other, then relay this information to a location processor.

Location Accuracy

RF trilateration uses estimated ranges from multiple receivers to estimate the location of a tag. RF triangulation uses the angles at which the RF signals arrive at multiple receivers to estimate the location of a tag. Many obstructions, such as walls or furniture, can distort the estimated range and angle readings leading to varied qualities of location estimate. Estimation-based locating is often measured in accuracy for a given distance, such as 90% accurate for 10 meter range.

Systems that use locating technologies that do not go through walls, such as infrared or ultrasound, tend to be more accurate in an indoor environment because only tags and receivers that have line of sight (or near line of sight) can communicate.

Applications

RTLS can be used numerous logistical or operational areas such as:

- locate and manage assets within a facility, such as finding a misplaced tool cart in a warehouse or medical equipment
- notification of new locations, such as an alert if a tool cart improperly has left the facility
- to combine identity of multiple items placed in a single location, such as on a pallet
- to locate customers, for example in a restaurant, for delivery of food or service
- to maintain proper staffing levels of operational areas, such as ensuring guards are in the proper locations in a correctional facility
- to quickly and automatically account for all staff after or during an emergency evacuation
- to automatically track and time stamp the progress of people or assets through a process, such as following a patient's emergency room wait time, time spent in the operating room, and total time until discharge. Such a system can be used for process improvement
- clinical-grade locating to support acute care capacity management

Privacy Concerns

RTLS may be seen as a threat to privacy when used to determine the location of people. The newly declared human right of informational self-determination de:Informationelle Selbstbestimmung gives the right to prevent one's identity and personal data from being disclosed to others, and also covers disclosure of locality, though this does not generally apply to the workplace.

Several prominent labor unions have come out against the use of RTLS systems to track workers calling them "the beginning of Big Brother" and "an invasion of privacy". A common strategy to overcome opposition to these systems is often similar to the following. However, this loss of privacy may be outweighed by other benefits to staff. For example, Toronto General Hospital is looking at RTLS to reduce quarantine times after an infectious disease outbreak. After a recent SARS outbreak, 1% of all staff were quarantined, and more accurate data regarding who had been exposed to the virus could have reduced the need for quarantines.The statement is very likely correct, but with say a staff of 1,000 10 people would require monitoring. Even if 50% could be eliminated, 5 persons from the quarantine which averaged 10 Days, it hardly seems worth it i the long run, especially considering how rare quarantines actually are.

Not to mention these issues are a slippery slope than need to be studied very carefully before being implemented. After all there is little doubt the government would also like this data so would be monitoring the entire city be ok? The state, or entire country? After all it *is* for their benefit.

Types of Technologies Used

There is a wide variety of systems concepts and designs to provide real-time locating.

- Active radio frequency identification (Active RFID)

- Active radio frequency identification - infrared hybrid (Active RFID-IR)
- Infrared (IR)
- Optical locating
- Low-frequency signpost identification
- Semi-active radio frequency identification (semi-active RFID)
- Passive RFID RTLS locating via Steerable Phased Array Antennae
- Radio beacon,
- Ultrasound Identification (US-ID)
- Ultrasonic ranging (US-RTLS)
- Ultra-wideband (UWB)
- Wide-over-narrow band
- Wireless Local Area Network (WLAN, Wi-Fi)
- Bluetooth,
- Clustering in noisy ambience,
- Bivalent systems

A general model for selection of the best solution for a locating problem has been constructed at the Radboud University of Nijmegen. Many of these references do not comply with the definitions given in international standardization with ISO/IEC 19762-5 and ISO/IEC 24730-1. However, some aspects of real-time performance are served and aspects of locating are addressed in context of absolute coordinates.

Ranging and Angulating

Depending on the physical technology used, at least one and often some combination of ranging and/or angulating methods are used to determine location:

- Angle of arrival (AoA)
- Line-of-sight (LoS)
- Time of arrival (ToA)
- Multilateration (Time difference of arrival) (TDoA)
- Time-of-flight (ToF)
- Two-way ranging (TWR) according to Nanotron's patents
- Symmetrical Double Sided – Two Way Ranging (SDS-TWR)
- Near-field electromagnetic ranging (NFER)

Errors And Accuracy

Real-time locating is affected by a variety of errors. Many of the major reasons relate to the physics of the locating system, and may not be reduced by improving the technical equipment.

None or no direct response

Many RTLS systems require direct and clear line of sight visibility. For those systems, where there is no visibility from mobile tags to fixed nodes there will be no result or a non valid result from locating engine. This applies to satellite locating as well as other RTLS systems such as angle of arrival and time of arrival. Fingerprinting is a way to overcome the visibility issue: If the locations in the tracking area contain distinct measurement fingerprints, line of sight is not necessarily needed. For example, if each location contains a unique combination of signal strength readings from transmitters, the location system will function properly. This is true, for example, with some Wi-Fi based RTLS solutions. However, having distinct signal strength fingerprints in each location typically requires a fairly high saturation of transmitters.

False location

The measured location may appear entirely faulty. This is a generally result of simple operational models to compensate for the plurality of error sources. It proves impossible to serve proper location after ignoring the errors.

Locating backlog

Real time is no registered branding and has no inherent quality. A variety of offers sails under this term. As motion causes location changes, inevitably the latency time to compute a new location may be dominant with regard to motion. Either an RTLS system that requires waiting for new results is not worth the money or the operational concept that asks for faster location updates does not comply with the chosen systems approach.

Temporary location error

Location will never be reported *exactly*, as the term *real-time* and the term *precision* directly contradict in aspects of measurement theory as well as the term *precision* and the term *cost* contradict in aspects of economy. That is no exclusion of precision, but the limitations with higher speed are inevitable.

Steady location error

Recognizing a reported location steadily apart from physical presence generally indicates the problem of insufficient over-determination and missing of visibility along at least one link from resident anchors to mobile transponders. Such effect is caused also by insufficient concepts to compensate for calibration needs.

Location jitter

Noise from various sources has an erratic influence on stability of results. The aim to provide a steady appearance increases the latency contradicting to real time requirements.

Location jump

As objects containing mass have limitations to jump, such effects are mostly beyond physical reality. Jumps of reported location not visible with the object itself generally indicate improper modeling with the location engine. Such effect is caused by changing dominance of various secondary responses.

Location creep

Location of residing objects gets reported moving, as soon as the measures taken are biased by secondary path reflections with increasing weight over time. Such effect is caused by simple averaging and the effect indicates insufficient discrimination of first echoes.

Standards

ISO/IEC

The basic issues of RTLS are standardized by the International Organization for Standardization and the International Electrotechnical Commission, under the ISO/IEC 24730 series. In this series of standards, the basic standard ISO/IEC 24730-1 identifies the terms describing a form of RTLS used by a set of vendors, but does not encompass the full scope of RTLS technology.

Currently several standards are published:

- ISO/IEC 19762-5:2008 Information technology — Automatic identification and data capture (AIDC) techniques — Harmonized vocabulary—Part 5: Locating systems
- ISO/IEC 24730-1:2014 Information technology — Real-time locating systems (RTLS) — Part 1: Application programming interface (API)
- ISO/IEC 24730-2:2012 Information technology — Real time locating systems (RTLS) — Part 2: Direct Sequence Spread Spectrum (DSSS) 2,4 GHz air interface protocol
- ISO/IEC 24730-5:2010 Information technology — Real-time locating systems (RTLS) — Part 5: Chirp spread spectrum (CSS) at 2,4 GHz air interface
- ISO/IEC 24730-21:2012 Information technology — Real time locating systems (RTLS) — Part 21: Direct Sequence Spread Spectrum (DSSS) 2,4 GHz air interface protocol: Transmitters operating with a single spread code and employing a DBPSK data encoding and BPSK spreading scheme
- ISO/IEC 24730-22:2012 Information technology — Real time locating systems (RTLS) — Part 22: Direct Sequence Spread Spectrum (DSSS) 2,4 GHz air interface protocol: Transmitters operating with multiple spread codes and employing a QPSK data encoding and Walsh offset QPSK (WOQPSK) spreading scheme
- ISO/IEC 24730-61:2013 Information technology — Real time locating systems (RTLS) — Part 61: Low rate pulse repetition frequency Ultra Wide Band (UWB) air interface
- ISO/IEC 24730-62:2013 Information technology — Real time locating systems (RTLS) —

Part 62: High rate pulse repetition frequency Ultra Wide Band (UWB) air interface

These standards do not stipulate any special method of computing locations, nor the method of measuring locations. This may be defined in specifications for trilateration, triangulation or any hybrid approaches to trigonometric computing for planar or spherical models of a terrestrial area.

INCITS

- INCITS 371.1:2003, Information Technology - Real Time Locating Systems (RTLS) - Part 1: 2.4 GHz Air Interface Protocol
- INCITS 371.2:2003, Information Technology - Real Time Locating Systems (RTLS) - Part 2: 433-MHz Air Interface Protocol
- INCITS 371.3:2003, Information Technology - Real Time Locating Systems (RTLS) - Part 3: Application Programming Interface

Limitations and Further Discussion

In RTLS application in the Healthcare industry, various studies were issued discussing the limitations of the currently adopted RTLS. Currently used technologies RFID, Wi-fi, UWB, all RFID based are hazardous in the sense of interference with sensitive equipment. A study carried out by Dr Erik Jan van Lieshout of the Academic Medical Centre of the University of Amsterdam published in 'JAMA' claimed "RFID and UWB could shut down equipment patients rely on" as "RFID caused interference in 34 of the 123 tests they performed". The first Bluetooth RTLS provider in the medical industry is supporting this in their article: "The fact that RFID cannot be used near sensitive equipment should in itself be a red flag to the medical industry". The RFID Journal responded to this study not negating it rather explaining real-case solution: "The Purdue study showed no effect when ultrahigh-frequency (UHF) systems were kept at a reasonable distance from medical equipment. So placing readers in utility rooms, near elevators and above doors between hospital wings or departments to track assets is not a problem". However the case of 'keeping at a reasonable distance' might be still an open question for the RTLS technology adopters and providers in medical facilities.

In many applications it is very difficult and at the same time important to make a proper choice among various communication technologies (e.g., RFID, WiFi, etc.) which RTLS may include. Wrong design decision made at early stages can lead to catastrophic results fore the system and a significant loss of money for fixing and redesign. To solve this problem a special metodology for RTLS design space exploration was developed. It consists of such steps as modelling, requirements specification and verification into a single efficient process.

Microchip Implant (Animal)

A microchip implant is an identifying integrated circuit placed under the skin of an animal. The chip, about the size of a large grain of rice, uses passive RFID (Radio Frequency Identification) technology, and is also known as a PIT tag (for Passive Integrated Transponder).

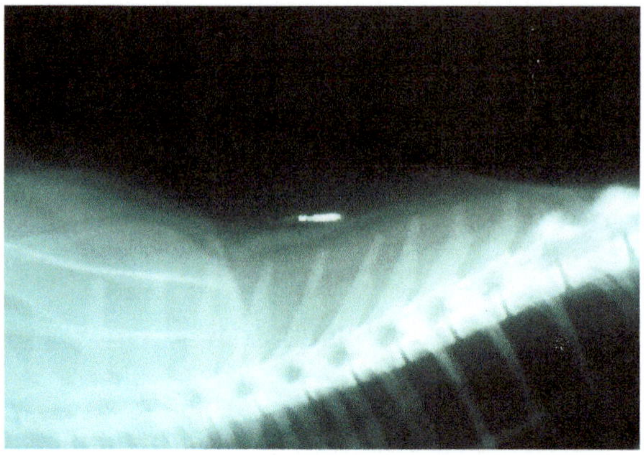

Microchip implant in a cat.

Externally attached microchips such as RFID ear tags are commonly used to identify farm and ranch animals other than horses. Some external microchips can be read with the same scanner used with implanted chips.

Uses and Benefits

Animal shelters, animal control officers and veterinarians routinely look for microchips to return lost pets quickly to their owners, avoiding expenses for housing, food, medical care, outplacing and euthanasia. Many shelters place chips in all outplaced animals.

Microchips are also used by kennels, breeders, brokers, trainers, registries, rescue groups, humane societies, clinics, farms, stables, animal clubs and associations, researchers, and pet stores.

Microchips can activate some pet doors programmed to recognize specific animals.

Some countries require microchips in imported animals to match vaccination records. Microchip tagging may also be required for CITES-regulated international trade in certain endangered animals: for example, Asian Arowana are tagged to limit import to captive-bred fish. Also, birds not banded who cross international borders as pets or for trade must be microchipped so that each bird is uniquely identifiable.

Usage

Information about the implant is often imprinted on a collar tag worn by a pet

Microchips can be implanted by a veterinarian or at a shelter. After checking that the animal does not already have a chip, the vet or technician injects the chip with a syringe and records the chip's unique ID. No anesthetic is required. A test scan ensures correct operation.

An enrollment form is completed with chip ID, owner contact information, pet name and description, shelter and/or veterinarian contact information, and an alternate emergency contact designated by the pet owner. Some shelters and vets designate themselves as the primary contact to remain informed about possible problems with the animals they place. The form is sent to a registry, who may be the chip manufacturer, distributor or an independent entity such as a pet recovery service. Some countries have a single official national database. For a fee, the registry typically provides 24-hour, toll-free telephone service for the life of the pet. Some veterinarians leave registration to the owner, usually done online, but a chip without current contact information is essentially useless.

The owner receives a registration certificate with the chip ID and recovery service contact information. The information can also be imprinted on a collar tag worn by the animal. Like an automobile title, the certificate serves as proof of ownership and is transferred with the animal when it is sold or traded; an animal without a certificate could be stolen.

Authorities and shelters examine strays for chips, providing the recovery service with the ID number, description and location so they may notify the owner or contact. If the pet is wearing the collar tag, the finder does not need a chip reader to contact the registry. An owner can also report a missing pet to the recovery service, as vets look for chips in new animals and check with the recovery service to see if it has been reported lost or stolen.

Many veterinarians scan an animal's chip on every visit to verify correct operation. Some use the chip ID as their database index and print it on receipts, test results, vaccination certifications and other records.

Components of a Microchip

A microchip implant is a passive RFID device. Lacking an internal power source, it remains inert until it is powered by the scanner.

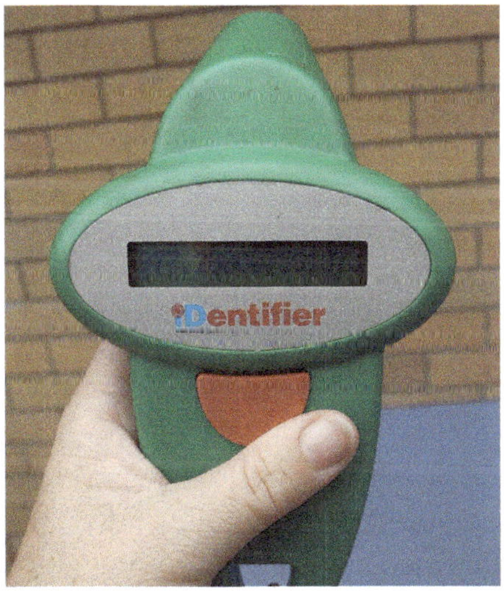

Example of an RFID scanner used with animal microchip implants.

Most implants contain three elements: a 'chip' or integrated circuit; a coil inductor, possibly with a ferrite core; and a capacitor. The chip contains unique identification data and electronic circuits to encode that information. The coil acts as the secondary winding of a transformer, receiving power inductively coupled to it from the scanner. The coil and capacitor together form a resonant LC circuit tuned to the frequency of the scanner's oscillating magnetic field to produce power for the chip. The chip then transmits its data back through the coil to the scanner.

These components are encased in biocompatible soda lime or borosilicate glass and hermetically sealed. Barring rare complications, dogs and cats are unaffected by them.

Implant Location

In dogs and cats, chips are usually inserted below the skin at the back of the neck between the shoulder blades on the dorsal midline. According to one reference, continental European pets get the implant in the left side of the neck. The chip can often be felt under the skin. Thin layers of connective tissue form around the implant and hold it in place.

Horses are microchipped on the left side of the neck, halfway between the poll and withers and approximately one inch below the midline of the mane, into the nuchal ligament.

Birds are implanted in their breast muscles. Proper restraint is necessary so the operation requires either two people (an avian veterinarian and a veterinary technician) or general anesthesia.

Implanted microchips can distort magnetic resonance imaging (MRIs), including those of the spinal cord.

Animal Species

Horse microchipping

Many animal species have been microchipped, including cockatiels and other parrots, horses, llamas, alpacas, goats, sheep, miniature pigs, rabbits, deer, ferrets, penguins, sharks, snakes, lizards, alligators, turtles, toads, frogs, rare fish, chimpanzees, mice, and prairie dogs—even whales and elephants. The U.S. Fish and Wildlife Service uses microchipping in its research of wild bison, black-footed ferrets, grizzly bears, elk, white-tailed deer, giant land tortoises and armadillos.

Worldwide Use

Microchips are not yet universal, but they are legally required in some jurisdictions such as the state of New South Wales, Australia and the United Kingdom (since 6 April 2016).

Some countries, such as Japan, require ISO-compliant microchips or a compatible reader on imported dogs and cats.

In New Zealand, all dogs first registered after 1 July 2006 must be microchipped. Farmers protested that farm dogs should be exempt, drawing a parallel to the Dog Tax War of 1898. Farm dogs were exempted from microchipping in an amendment to the legislation passed in June 2006. A National Animal Identification and Tracing scheme in New Zealand is currently being developed for tracking livestock.

In April 2012 Northern Ireland became the first part of the United Kingdom to require microchipping of individually licensed dogs. Dog microchipping became mandatory in England on 6 April 2016.

In Israel, microchips in dogs are mandatory.

Australia has a National Livestock Identification System.

The United States uses the National Animal Identification System for farm and ranch animals other than dogs and cats. In most species except horses, an external eartag is typically used in lieu of an implant microchip. Eartags with microchips or simply stamped with a visible number can be used. Both use ISO 15 digit microchip numbers with the U.S. country code of 840.

Cross-compatibility and Standards Issues

In most countries, pet ID chips adhere to an international standard to promote compatibility between chips and scanners. In the United States, however, three proprietary types of chips compete along with the international standard. Scanners distributed to United States shelters and veterinarians well into 2006 could each read at most three of the four types. Scanners with quad-read capability are now available and are increasingly considered required equipment. Older scanner models will be in use for some time, so United States pet owners must still choose between a chip with good coverage by existing scanners and one compatible with the international standard. The four types include:

- The ISO Conformant Full Duplex type has the greatest international acceptance. It is common in many countries including Europe (since the late 1990s) and Canada. It is one of two chip protocol types (along with the "Half Duplex" type sometimes used in farm and ranch animals) that conform to International Organization for Standardization standards ISO 11784 & 11785. To support international/multivendor application, the 3-digit country code can contain an assigned ISO country code or a manufacturer code from 900 to 998 plus its identifying serial number. In the United States, distribution of this type has been controversial. When 24PetWatch.com began distributing them in 2003 (and more famously Banfield Pet Hospitals in 2004) many shelter scanners couldn't read them. (Some still can't; asking local shelters about this is still a good idea.) At least one Banfield-chipped pet was inadvertently euthanized.

- The Trovan Unique type is another pet chip protocol type in use since 1990 in pets in the United States. Patent problems forced the withdrawal of Trovan's implanter device from United States distribution and they became uncommon in pets in the United States, although Trovan's original registry database "infopet.biz" remained in operation. In early 2007, the American Kennel Club's chip registration service, AKC Companion Animal Recovery Corp, which had been the authorized registry for HomeAgain brand chips made by Destron/Digital Angel, began distributing Trovan chips with a different implanter. These chips are read by the Trovan, HomeAgain (Destron Fearing), and Bayer (Black Label) readers. Despite multiple offers from Trovan to AVID to license the technology to read the Trovan chips, AVID continues to distribute readers that do not read Trovan or the ISO compliant chips.
- A third type sometimes known as FECAVA or Destron is available under various brand names. These include, in the United States, "Avid Eurochip", the common current 24PetWatch chips, and the original (and still popular) style of HomeAgain chips. (HomeAgain and 24Petwatch can now supply the true ISO chip instead on request.) Chips of this type have 10 digit [hexadecimal] chip numbers. This "FECAVA" type is readable on a wide variety of scanners in the United States and has been less controversial, although its level of adherence to the ISO standards is sometimes exaggerated in some descriptions. The ISO standard has an annex (appendix) recommending that three older chip types be supported by scanners, including a 35-bit "FECAVA"/"Destron" type. The common Eurochip/HomeAgain chips don't agree perfectly with the annex description, although the differences are sometimes considered minor. But the ISO standard also makes it clear that only its 64-bit "full-duplex" and "half-duplex" types are "conformant"; even chips (e.g., the Trovan Unique) that match one of the Annex descriptions are not. More visibly, FECAVA cannot support the ISO standard's required country/manufacturer codes. They may be accepted by authorities in many countries where ISO-standard chips are the norm, but not by those requiring literal ISO conformance.
- Finally, there's the AVID brand Friendchip type, which is peculiar due to its encryption characteristics. Cryptographic features are not necessarily unwelcome; few pet rescuers or humane societies would object to a design that outputs an ID number "in the clear" for anyone to read, along with authentication features for detection of counterfeit chips, but the authentication in "Friendchips" has been found lacking and rather easy to spoof to the AVID scanner. Although no authentication encryption is involved, obfuscation requires proprietary information to convert transmitted chip data to its original label ID code. Well into 2006, scanners containing the proprietary decryption were provided to the United States market only by AVID and Destron/Digital Angel; Destron/Digital Angel put the decryption feature in some, but not all, of its scanners, possibly as early as 1996. (For years, its scanners distributed to shelters through HomeAgain usually had full decryption, while many sold to veterinarians would only state that an AVID chip had been found.) Well into 2006, both were resisting calls from consumers and welfare group officials to bring scanners to the United States shelter community combining AVID decryption capability with the ability to read ISO-compliant chips. Some complained that AVID itself had long marketed combination pet scanners compatible with all common pet chips except possibly Trovan outside the United States. By keeping them out of the United States, it could be

considered partly culpable in the missed-ISO chips problem others blamed on Banfield.[Note 5] In 2006, the European manufacturer Datamars, a supplier of ISO chips used by Banfield and others, gained access to the decryption secrets and began supplying scanners with them to United States customers. This "Black Label" scanner was the first four-standard full-multi pet scanner in the United States market. Later in 2006, Digital Angel announced that it would supply a full-multi scanner in the United States. In 2008 AVID announced a "breakthrough" scanner, although as of October 2010 AVID's is still so uncommon that it's unclear whether it supports the Trovan chip. Trovan also acquired the decryption technology in 2006 or earlier, and now provides it in scanners distributed in the United States by AKC-CAR. (Some are quad-read, but others lack full ISO support.)

Numerous references in print state that the incompatibilities between different chip types are a matter of "frequency". One may find claims that early ISO adopters in the United States endangered their customers' pets by giving them ISO chips that work at a "different frequency" from the local shelter's scanner, or that the United States government considered forcing an incompatible frequency change. These claims were little challenged by manufacturers and distributors of ISO chips, although later evidence suggests the claims were disinformation. In fact, all chips operate at the scanner's frequency. Although ISO chips are optimized for 134.2 kHz, in practice they are readable at 125 kHz and the "125 kHz" chips are readable at 134.2 kHz. Confirmation comes from government filings that indicate the supposed "multi-frequency" scanners now commonly available are really single-frequency scanners operating at 125, 134.2 or 128 kHz. In particular, the United States HomeAgain scanner didn't change excitation frequency when ISO-read capability was added; it's still a single frequency, 125 kHz scanner.

Scanner Compatibility table for chip types used in pets				
	Expected results for chip type (OK=Good read NR=No read DO=Detect Only with no number given)			
Scanner to test	ISO Conformant Full Duplex chip	AVID Encrypted "FriendChip"	Original U.S. HomeAgain, AVID Eurochip, or FECAVA	"Trovan Unique" and current AKC CAR chips
Minimal ISO Conformant Scanner (also must read HALF Duplex chips common in livestock ear tags)	OK	NR	NR	NR
AVID Basic U.S. Scanner	NR	OK	NR	NR
AVID Deluxe U.S. Scanner	NR	OK	OK	NR
AVID Universal Scanner sold outside U.S.	OK	OK	OK	NR Assumed
AVID MiniTracker Pro Scanner announced August 2008	OK	OK	OK	NR according to some (Few have seen one.)

Various vintages of U.S. HomeAgain "Universal" Shelter Scanners by Destron/Digital Angel Corp.	NR,DO, or OK	OK	OK	Possibly all OK
Typical Destron/ Digital Angel Corp. U.S. Veterinarian's scanner pre-2007	NR	DO	OK	DO
Trovan LID-560-MULTI per mfr. specifications on Web	OK	OK	OK	OK
U.S. Trovan Pocket Scanner per AKC-CAR Web Site	DO	OK	OK	OK
U.S. Trovan ProScan700 per AKC-CAR Web Site	OK	OK	OK	OK
Original 2006 Datamars Black Label Scanner	OK	OK	OK	OK but Reliability Questioned
Datamars Black Label Scanner "classypets" model	OK	NR or DO?	OK	OK but Reliability Questioned
Banfield-Distributed 2004-2005 Vintage Datamars Scanners	OK	Possibly all DO	OK	Possibly all OK but Reliability Questioned (Undocumented Feature)
Datamars Minimax and Micromax	OK	NR	NR	NR
Typical Homemade Scanner	OK	OK but extra step required (web-based decryption service)	OK	OK

(For users requiring Shelter-Grade certainty, this table is not a substitute for testing the scanner with a set of specimen chips. One study cites problems with certain Trovan chips on the Datamars Black Label scanner. In general, the study found none of the tested scanners to read all four standards without some deficiency. The study predates the most recent scanner models, however.)

Reported Adverse Reactions

RFID chips are used in animal research, and at least three studies conducted since the 1990s have reported tumors at the site of implantation in laboratory mice and rats. Noted veterinary associations responded with continued support for the procedure as reasonably safe for cats and dogs, pointing to rates of serious complications on the order of one in a million in the U.K., which has a system for tracking such adverse reactions and has chipped over 3.7 million pet dogs. A recent study found no safety concerns for microchipped animals with RFID chips undergoing MRI at one Tesla magnetic field strength. In 2011 a microchip-associated fibrosarcoma was reported found in the neck of a 9-year old, neutered-male cat. Histological examination was consistent with postinjection sarcoma, but all prior vaccinations occurred in the hindlegs.

Transponder Timing

A finish line which uses transponder timing and RFID technology through overhead antennas and passive, disposable chips.

Transponder timing (also called chip timing or RFID timing) is a technique for measuring performance in sport events. A transponder working on a radio-frequency identification (RFID) basis is attached to the athlete and emits a unique code that is detected by radio receivers located at the strategic points in an event.

Prior to the use of this technology, races were either timed by hand (with operators pressing a stopwatch) or using video camera systems.

Transponder Systems

Generically, there are two types of transponder timing systems; active and passive. An active transponder consists of a battery-powered transceiver, connected to the athlete, that emits its unique code when it is interrogated.

A passive transponder does not contain a power source inside the transponder. Instead, the transponder captures electromagnetic energy produced by a nearby exciter and utilizes that energy to emit a unique code.

In both systems, an antenna is placed at the start, finish, and in some cases, intermediate time points and is connected to a decoder. This decoder identifies the unique transponder code and calculates the exact time when the transponder passes a timing point. Some implementations of timing systems require the use of a mat on the ground at the timing points while other systems implement the timing points with vertically oriented portals.

History

RFID was first used in the late 1980s primarily for motor racing and became more widely adopted in athletic events in the mid-1990s on release of low cost 134 kHz transponders and readers from Texas Instruments. This technology formed the basis of electronic sports timing for the world's largest running events as well as for cycling, triathlon and skiing. Some manufacturers made improvements to the technology to handle larger numbers of transponders in the read field or improve the tolerance of their systems to low frequency noise. These low frequency systems are still used a lot today. Other manufacturers developed their own proprietary RFID systems usually as an offshoot to more industrial applications. These latter systems attempted to get around the problem of reading large numbers of transponders in a read field by using the High Frequency 13.56 MHz RFID methodology that allowed transponders to use anti-collision algorithms to avoid tags interfering with each other's signal during the down-link between transponder and reader. Active transponder systems continued to mature and despite their much higher cost they retained market share in the high speed sports like motor racing, cycling and ice skating. Active systems are also used at high profile events such as the Olympics due to their very high read rates and time-stamping precision.

By 2005 a newer RFID technology was becoming available, mostly for industrial applications. The first and second generation (UHF) transponders and readers that were being developed followed a strict protocol to ensure that multiple transponders and readers could be used between manufacturers. Much like the HF tags, the UHF tags were much cheaper to produce in volume and formed the basis in the next revolution in sports timing. Currently many of the largest athletic events are timed using disposable transponders either placed on the back of a race number or on the runners shoe. The low cost meant that transponders were now fully disposable and did not need to be returned to the organizers after the event.

Usage

Very large running events (more than 10,000) and triathlons were the first events to be transponder (or chip) timed because it is near impossible to manually time them. Also for large runs there are delays in participants reaching the start line, which penalize their performance. Some races place antennas or timing mats at both the start line and the finish line, which allow the exact net time to be calculated. Awards in a race are generally based on the "gun time" (which ignores any delay at the start) as per IAAF and USA Track and Field rules. However, some races use "net time" for presenting age group awards.

In the past the transponder was almost always worn on the athletes running shoe, or on an ankle band. This enabled the transponder to be read best on antenna mats because the distance between the transponder and readers antenna is minimized offering the best capture rate. Transponders may be threaded onto the shoe laces for running. For triathlon a soft elastic ankle band holds the transponder to the leg and care is taken to ensure the transponder is in the correct orientation or polarity for maximum read performance. Transponders have also been placed on the race bib. In the past 5 years[*when?*] the newer UHF systems use transponders placed on the shoe lace, or stuck to the race number bib. In both cases, care must be taken to ensure the UHF tag does not directly touch a large part of the skin as this affects read performance. Despite this, UHF Systems have read performances as good (if not better) than the conventional low and high frequency systems. Because these UHF tags are made in huge volumes for industrial applications, their price is much lower than that of conventional re-usable transponders and the race does not bother to collect them afterwards.

Disposable bib with two passive timing chips at the back

Back side of disposable RFID tag used for race timing.

All RFID timing systems incorporate a box housing the reader(s) with peripherals like a microprocessor, serial or ethernet communications and power source (battery). The readers are attached to one or more antennas that are designed for the particular operating frequency. In the case of low or medium frequencies these consist of wire loops incorporated into mats that cover the entire width of the timing point. For UHF systems the antennas consist of patch antennas that are protected in a matting system. The patch antennas may also be placed on stands or a finish gantry pointing towards the oncoming athlete. In most cases the distance between reader and antennas is restricted. Also more equipment is needed for events that require multiple timing points. Wider timing points require more readers and antennas. For active systems a simple wire loop is all that is needed since the transponder has its own power source and the loop serves as a trigger to turn on the transponder, then receive the relatively strong signal from the transponder. Therefore, active systems need less readers (or decoders) per timing point width.

All systems utilize specialised software to calculate results and splits. This software usually resides on a separate PC computer that is connected to the readers via serial or ethernet communications. The software relates the raw transponder code and timestamp data to each entrant in a database and calculates gun and net times of runners, or the splits of a triathlete. In advanced systems these results are instantly calculated and published to the internet so that athletes and spectators have access to results via any web enabled device.

References

- Hacking Exposed Linux: Linux Security Secrets & Solutions (third ed.). McGraw-Hill Osborne Media. 2008. p. 298. ISBN 978-0-07-226257-5.
- Sen, Dipankar; Sen, Prosenjit; Das, Anand M. (2009), RFID For Energy and Utility Industries, PennWell, ISBN 978-1-59370-105-5, pp. 1-48
- Ilyas, edited by Syed Ahson, Mohammad (2008). "26.5". RFID handbook : applications, technology, security, and privacy ([Online-Ausg.] ed.). Boca Raton: CRC Press. p. 478. ISBN 9781420054996. Retrieved 7 August 2012.
- Katherine Albrecht; Liz McIntyre (2005). Spychips: how major corporations and government plan to track your every move with RFID. Thomas Nelson Inc. ISBN 1-59555-020-8.
- Coren, Michael J. (2011-12-05). "VA's Real-Time Location System: A way to improve patient safety, or Big Brother?". Nextgov.com. Retrieved 2016-04-28.
- "How RF Controls Technology Paves the Way for the "Internet of Everything." | RF Controls". Rfctrls.com. 2014-05-07. Retrieved 2016-04-28.
- "RFID Technology from Texas Instruments and RF Code Brings Service and Safety to Guests at Steamboat Ski Resort" (PDF). Rfidjournalevents.com. Retrieved 2016-04-28.
- "IEEE Xplore Abstract - Enhancing Accuracy Performance of Bluetooth Positioning". Ieeexplore.ieee.org. 2007-03-15. doi:10.1109/WCNC.2007.506. Retrieved 2016-04-28.
- "IEEE Xplore Abstract - WLAN location determination via clustering and probability distributions". Ieeexplore.ieee.org. 2003-03-26. doi:10.1109/PERCOM.2003.1192736. Retrieved 2016-04-28.
- "JAMA Network | JAMA | Electromagnetic Interference From Radio Frequency Identification Inducing Potentially Hazardous Incidents in Critical Care Medical Equipment". Jama.jamanetwork.com. Retrieved 2016-04-28.
- "ISO/IEC 19762-5:2008 - Information technology - Automatic identification and data capture (AIDC) techniques - Harmonized vocabulary - Part 5: Locating systems". Iso.org. Retrieved 2016-04-28.

5

Wireless Technologies and Applications

Radio navigation is an application that is used to trace an object on Earth whereas radars are used to determine the position and range of an object. Countries use radars to detect unknown aircrafts, ships, spacecrafts and guided missiles. The other wireless technologies and applications explained in this chapter are Wi-Fi, hotspot and Li-Fi.

Radio Navigation

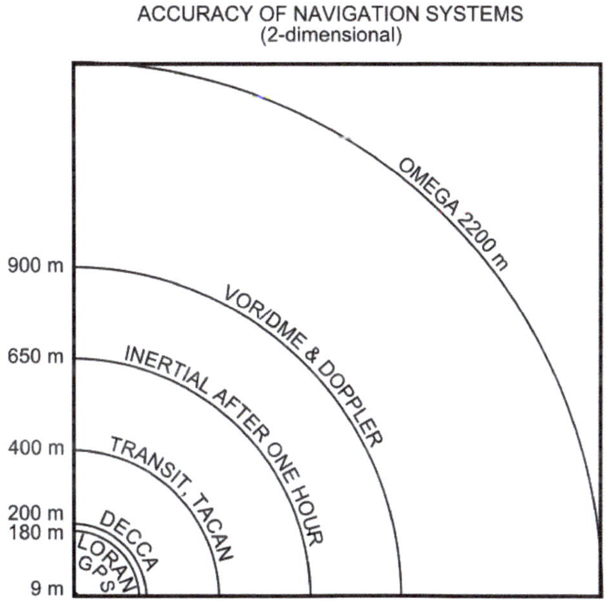

Radio navigation or radionavigation is the application of radio frequencies to determine a position of an object on the Earth. Like radiolocation, it is a type of radiodetermination.

The basic principles are measurements from/to electric beacons, especially

- directions, e.g. by bearing, radio phases or interferometry,
- distances, e.g. ranging by measurement of travel times,
- partly also velocity, e.g. by means of radio Doppler shift.

Bearing-measurement Systems

These systems used some form of directional radio antenna to determine the location of a broadcast station on the ground. Conventional navigation techniques are then used to take a radio fix. These were introduced prior to WWI, and remain in use today.

Radio Direction Finding

Amelia Earhart's Lockheed Electra had a prominent RDF loop on the cockpit roof.

The first system of radio navigation was the *Radio Direction Finder*, or RDF. By tuning in a radio station and then using a directional antenna, one could determine the direction to the broadcasting antenna. A second measurement using another station was then taken. Using triangulation, the two directions can be plotted on a map where their intersection reveals the location of the navigator. Commercial AM radio stations can be used for this task due to their long range and high power, but strings of low-power radio beacons were also set up specifically for this task, especially near airports and harbours.

Early RDF systems normally used a loop antenna, a small loop of metal wire that is mounted so it can be rotated around a vertical axis. At most angles the loop has a fairly flat reception pattern, but when it is aligned perpendicular to the station the signal received on one side of the loop cancels the signal in the other, producing a sharp drop in reception known as the "null". By rotating the loop and looking for the angle of the null, the relative bearing of the station can be determined. Loop antennas can be seen on most pre-1950s aircraft and ships.

Reverse RDF

The Orfordness Beacon as it appears today.

The main problem with RDF is that it required a special antenna on the vehicle, which may not be easy to mount on smaller vehicles or single-crew aircraft. A smaller problem is that the accuracy of the system is based to a degree on the size of the antenna, but larger antennas would likewise make the installation more difficult.

During the era between World War I and World War II, a number of systems were introduced that placed the rotating antenna on the ground. As the antenna rotated through a fixed position, typically due north, the antenna was keyed with the morse code signal of the station's identification letters so the receiver could ensure they were listening to the right station. Then they waited for the signal to either peak or disappear as the antenna briefly pointed in their direction. By timing the delay between the morse signal and the peak/null, then dividing by the known rotational rate of the station, the bearing of the station could be calculated.

The first such system was the German Telefunken Kompass Sender, which began operations in 1907 and was used operationally by the Zeppelin fleet until 1918. An improved version was introduced by the UK as the Orfordness Beacon in 1929 and used until the mid-1930s. A number of improved versions followed, replacing the mechanical motion of the antennas with phasing techniques that produced the same output pattern with no moving parts. One of the longest lasting examples was Sonne, which went into operation just before World War II and used operationally under the name Consol until 1991. The modern VOR system is based on the same principles.

ADF and NDB

A great advance in the RDF technique was introduced in the form of phase comparisons of a signal as measured on two or more small antennas, or a single highly directional solenoid. These receivers were dramatically smaller, more accurate, and simpler to operate. Combined with the introduction of the transistor and integrated circuit, RDF systems were so reduced in size and complexity that they once again became quite common during the 1960s, and were known by the new name, automatic direction finder, or ADF.

This also led to a revival in the operation of simple radio beacons for use with these RDF systems, now referred to as *non-directional beacons* (NDB). As the LF/MF signals used by NDBs can follow the curvature of earth, NDB has a much greater range than VOR which travels only in *line of sight*. NDB can be categorized as *long range* or *short range* depending on their power. The frequency band allotted to non-directional beacons is 190–1750 kHz, but the same system can be used with any common AM-band commercial station.

VOR

VOR transmitter station

VHF omnidirectional range, or VOR, is an implementation of the reverse-RDF system, but one that is more accurate and able to be completely automated.

Instead of a single signal, the VOR transmitter sends out three signals – one is a simple voice channel that sends morse code to identify the station, another is a continuous signal sent in all directions, and the last is a signal that is rotated at 30 RPM. Like the Orfordness concept, the bearing of the station is measured by finding the rotating signal's peak or null. But instead of timing the signal, the rotating signal is changed in phase in synchronicity with its rotation, such that it is in-phase when pointed north, 90 degrees off when it points east, and so forth. By comparing the phase of the received signal with the one being broadcast omnidirectionally, the angle can be determined using simple electronics. This angle is then displayed in the cockpit of the aircraft, and can be used to take a fix just like the earlier RDF systems, although it is easier to use.

As VOR required two VHF receivers as well as a conventional radio for station identification, the system did not become popular until the era of miniaturized electronics, first with small tubes in the 1950s, and then transistorized systems in the 1960s. During this period it quickly took over from the older Radio Range system. The signals from the stations could be received anywhere, as opposed to the beams which were only broadcast in certain directions, so in theory the VOR system could be used for free navigation from any to any point. In practice, the older Radio Range procedures were so widely used and standardized that VOR was used to produce a similar set of airways that remain in use today.

The US military also introduced a VOR-like system known as TACAN. It differed from VOR primarily in its modulation system, adding a Lorentz-like signal to accurately define the center of the rotating beam and thereby improve accuracy. It requires five receiver channels and additional electronics, an expensive requirement when it was introduced.

Beam Systems

Beam systems broadcast narrow signals in the sky, and navigation is accomplished by keeping the aircraft centred in the beam. A number of stations are used to create an airway, with the navigator tuning in different stations along the direction of travel. These systems were common in the era when electronics were large and expensive, as they placed minimum requirements on the receivers – they were simply voice radio sets tuned to the selected frequencies. However, they did not provide navigation outside of the beams, and were thus less flexible in use. The rapid miniaturization of electronics during and after WWII made systems like VOR practical, and most beam systems rapidly disappeared.

Lorenz

In the post-WWI era, the Lorenz company of Germany developed a means of projecting two narrow radio signals with a slight overlap in the center. By broadcasting different audio signals in the two beams, the receiver could position themselves very accurately down the centreline by listening to the signal in their headphones. The system was accurate to less than a degree in some forms.

Originally known as "Ultrakurzwellen-Landefunkfeuer" (LFF), or simply "Leitstrahl" (guiding beam), little money was available to develop a network of stations. Deployment was instead led by the US, where it formed the basis of a wide-area navigation system through the 1930s and 40s. Development was restarted in Germany in the 1930s as a short-range system deployed at airports as a blind landing aid. Although there was some interest in deploying a medium-range system like the US LFF, deployment had not yet started when the beam system was combined with the Orfordness timing concepts to produce the highly accurate Sonne system. In all of these roles, the system was generically known simply as a "Lorenz beam".

In the immediate pre-WWII era the same concept was also developed as a blind-bombing system. This used very large antennas to provide the required accuracy at long distances (over England), and very powerful transmitters. Two such beams were used, crossing over the target to triangulate it. Bombers would enter one of the beams and use it for guidance until they heard the second one in a second radio receiver, using that signal to time the dropping of their bombs. The system was highly accurate, and the 'Battle of the Beams' broke out when United Kingdom intelligence services attempted, and then succeeded, in rendering the system useless through electronic warfare. Sonne, however, proved just as useful to the UK as Germany, and was left to operate unhindered throughout the war.

Low Frequency Radio Range

The low-frequency radio range (LFR, also other names) was the main navigation system used by aircraft for instrument flying in the 1930s and 1940s in the U.S. and other countries, until the advent of the VOR in the late 1940s. It was used for both en route navigation as well as instrument approaches.

The ground stations consisted of a set of four antennas that projected Lorenz beams in four cardinal directions. One of the beams was "keyed" with the morse code signal "A", dit-dah, with the second beam "N", dah-dit. Flying down the centreline produced a steady tone. The beams were pointed to the next station to produce a set of airways, allowing an aircraft to travel from airport to airport by following a selected set of stations. Effective course accuracy was about three degrees, which near the station provided sufficient safety margins for instrument approaches down to low minimums. At its peak deployment, there were nearly 400 LFR stations in the US.

Glide Path and the Localizer of ILS

The remaining widely used beam systems are glide path and the localizer of the *instrument landing system* (ILS). ILS uses a *localizer* to provide horizontal position, distance to the runway, and airport information, and *glide path* to provide vertical positioning. ILS can provide enough accuracy and redundancy to allow automated landings.

Transponder Systems

Positions can be determined with any two measures of angle or distance. The introduction of radar in the 1930s provided a way to directly determine the distance to an object even at long distances. Navigation systems based on these concepts soon appeared, and remained in widespread use until recently. Today they are used primarily for aviation, although GPS has largely supplanted this role.

Radar and Transponders

Early radar systems, like the UK's Chain Home, consisted of large transmitters and separate receivers. The transmitter periodically sends out a short pulse of a powerful radio signal, which is sent into space through broadcast antennas. When the signal reflects off a target, some of that signal is reflected back in the direction of the station, where it is received. The received signal is a tiny fraction of the broadcast power, and has to be powerfully amplified in order to be used.

The same signals are also sent over local electrical wiring to the operator's station, which is equipped with an oscilloscope. Electronics attached to the oscilloscope provides a signal that increases in voltage over a short period of time, a few microseconds. When sent to the X input of the oscilloscope, this causes a horizontal line to be displayed on the scope. This "sweep" is triggered by a signal tapped off the broadcaster, so the sweep begins when the pulse is sent. Amplified signals from the receiver are then sent to the Y input, where any received reflection causes the beam to move upward on the display. This causes a series of "blips" to appear along the horizontal axis, indicating reflected signals. By measuring the distance from the start of the sweep to the blip, which corresponds to the time between broadcast and reception, the distance to the object can be determined.

Soon after the introduction of radar, the radio transponder appeared. Transponders are a combination of receiver and transmitter whose operation is automated – upon reception of a particular signal, normally a pulse on a particular frequency, the transponder sends out a pulse in response, typically delayed by some very short time. Transponders were initially used as the basis for early IFF systems; aircraft with the proper transponder would appear on the display as part of the normal radar operation, but then the signal from the transponder would cause a second blip to appear a short time later. Single blips were enemies, double blips friendly.

Transponder-based distance-distance navigation systems have a significant advantage in terms of positional accuracy. Any radio signal spreads out over distance, forming the fan-like beams of the Lorenz signal, for instance. As the distance between the broadcaster and receiver grows, the area covered by the fan increases, decreasing the accuracy of location within it. In comparison, transponder-based systems measure the timing between two signals, and the accuracy of that measure is largely a function of the equipment and nothing else. This allows these systems to remain accurate over very long range.

The latest transponder systems (mode S) can also provide position information, possibly derived from GNSS, allowing for even more precise positioning of targets.

Bombing Systems

The first distance-based navigation system was the German Y-Gerät blind-bombing system. This used a Lorenz beam for horizontal positioning, and a transponder for ranging. A ground-based system periodically sent out pulses which the airborne transponder returned. By measuring the total round-trip time on a radar's oscilloscope, the aircraft's range could be accurately determined even at very long ranges. An operator then relayed this information to the bomber crew over voice channels, and indicated when to drop the bombs.

The British introduced similar systems, notably the Oboe system. This used two stations in England that operated on different frequencies and allowed the aircraft to be triangulated in space. To ease pilot workload only one of these was used for navigation – prior to the mission a circle was drawn over the target from one of the stations, and the aircraft was directed to fly along this circle on instructions from the ground operator. The second station was used, as in Y-Gerät, to time the bomb drop. Unlike Y-Gerät, Oboe was deliberately built to offer very high accuracy, as good as 35 m, much better than even the best optical bombsights.

One problem with Oboe was that it allowed only one aircraft to be guided at a time. This was addressed in the later Gee-H system by placing the transponder on the ground and broadcaster in the aircraft. The signals were then examined on existing Gee display units in the aircraft. Gee-H did not offer the accuracy of Oboe, but could be used by as many as 90 aircraft at once. This basic concept has formed the basis of most distance measuring navigation systems to this day.

Beacons

The key to the transponder concept is that it can be used with existing radar systems. The ASV radar introduced by RAF Coastal Command was designed to track down submarines and ships by displaying the signal from two antennas side by side and allowing the operator to compare their relative strength. Adding a ground-based transponder immediately turned the same display into a system able to guide the aircraft towards a transponder, or "beacon" in this role, with high accuracy.

The British put this concept to use in their Rebecca/Eureka system, where battery-powered "Eureka" transponders were triggered by airborne "Rebecca" radios and then displayed on ASV Mk. II radar sets. Eureka's were provided to French resistance fighters, who used them to call in supply drops with high accuracy. The US quickly adopted the system for paratroop operations, dropping the Eureka with pathfinder forces or partisans, and then homing in on those signals to mark the drop zones.

The beacon system was widely used in the post-war era for blind bombing systems. Of particular note were systems used by the US Marines that allowed the signal to be delayed in such a way to offset the drop point. These systems allowed the troops at the front line to direct the aircraft to points in front of them, directing fire on the enemy. Beacons were widely used for temporary or mobile navigation as well, as the transponder systems were generally small and low-powered, able to be man portable or mounted on a Jeep.

DME

In the post-war era, a general navigation system using transponder-based systems was deployed as the distance measuring equipment (DME) system.

DME was identical to Gee-H in concept, but used new electronics to automatically measure the time delay and display it as a number, rather than having the operator time the signals manually on an oscilloscope. This led to the possibility that DME interrogation pulses from different aircraft might be confused, but this was solved by having each aircraft send out a different series of pulses which the ground-based transponder repeated back.

DME is almost always used in conjunction with VOR, and is normally co-located at a VOR station.

This combination allows a single VOR/DME station to provide both angle and distance, and thereby provide a single-station fix. DME is also used as the distance-measuring basis for the military TACAN system, and their DME signals can be used by civilian receivers.

Hyperbolic Systems

Hyperbolic navigation systems are a modified form of transponder systems which eliminate the need for an airborne transponder. The name refers to the fact that they do not produce a single distance or angle, but instead indicate a location along any number of hyperbolic lines in space. Two such measurements produces a fix. As these systems are almost always used with a specific navigational chart with the hyperbolic lines plotted on it, they generally reveal the receiver's location directly, eliminating the need for manual triangulation. As these charts were digitized, they became the first true location-indication navigational systems, outputting the location of the receiver as latitude and longitude. Hyperbolic systems were introduced during WWII and remained the main long-range advanced navigation systems until GPS replaced them in the 1990s.

Gee

The first hyperbolic system to be developed was the British Gee system, developed during World War II. Gee used a series of transmitters sending out precisely timed signals, with the signals leaving the stations at fixed delays. An aircraft using Gee, RAF Bomber Command's heavy bombers, examined the time of arrival on an oscilloscope at the navigator's station. If the signal from two stations arrived at the same time, the aircraft must be an equal distance from both transmitters, allowing the navigator to determine a line of position on his chart of all the positions at that distance from both stations. More typically, the signal from one station would be received earlier than the other. The *difference* in timing between the two signals would reveal them to be along a curve of possible locations. By making similar measurements with other stations, additional lines of position can be produced, leading to a fix. Gee was accurate to about 165 yards (150 m) at short ranges, and up to a mile (1.6 km) at longer ranges over Germany. Gee remained in use long after WWII, and equipped RAF aircraft as late as the 1960s (approx freq was by then 68 MHz).

LORAN

With Gee entering operation in 1942, similar US efforts were seen to be superfluous. They turned their development efforts towards a much longer-ranged system based on the same principles, using much lower frequencies that allowed coverage across the Atlantic Ocean. The result was LORAN, for "LOng-range Aid to Navigation". The downside to the long-wavelength approach was that accuracy was greatly reduced compared to the high-frequency Gee. LORAN was widely used during convoy operations in the late war period.

Decca

Another British system from the same era was Decca Navigator. This differed from Gee primarily in that the signals were not pulses delayed in time, but continuous signals delayed in phase. By comparing the phase of the two signals, the time difference information as Gee was returned. However, this was far easier to display; the system could output the phase angle to a pointer on a dial removing any need for visual interpretation. As the circuitry for driving this display was quite small, Decca systems normally used three such displays, allowing quick and accurate reading of multiple fixes. Decca found its greatest use post-war on ships, and remained in use into the 1990s.

LORAN-C

Almost immediately after the introduction of LORAN, in 1952 work started on a greatly improved version. LORAN-C (the original retroactively became LORAN-A) combined the techniques of pulse timing in Gee with the phase comparison of Decca.

The resulting system (operating in the low frequency (LF) radio spectrum from 90 to 110 kHz) that was both long-ranged (for 60 kW stations, up to 3400 miles) and accurate. To do this, LORAN-C sent a pulsed signal, but modulated the pulses with an AM signal within it. Gross positioning was determined using the same methods as Gee, locating the receiver within a wide area. Finer accuracy was then provided by measuring the phase difference of the signals, overlaying that second measure on the first. By 1962, high-power LORAN-C was in place in at least 15 countries.

LORAN-C was fairly complex to use, requiring a room of equipment to pull out the different signals. However, with the introduction of integrated circuits, this was quickly reduced further and further. By the late 1970s LORAN-C units were the size of a stereo amplifier and were commonly found on almost all commercial ships as well as some larger aircraft. By the 1980s this had been further reduced to the size of a conventional radio, and it became common even on pleasure boats and personal aircraft. It was the most popular navigation system in use through the 1980s and 90s, and its popularity led to many older systems being shut down, like Gee and Decca. However, like the beam systems before it, civilian use of LORAN-C was short-lived when GPS technology drove it from the market.

Other Hyperbolic Systems

Similar hyperbolic systems included the US global-wide VLF/Omega Navigation System, and the similar Alpha deployed by the USSR. These systems determined pulse timing not by comparison of two signals, but by comparison of a single signal with a local atomic clock. The expensive-to-maintain Omega system was shut down in 1997 as the US military migrated to using GPS. Alpha is still in use.

Satellite Navigation

Since the 1960s, navigation has increasingly moved to satellite navigation systems. These are essentially DME systems located in space. The fact that the satellites are in orbit and normally move with respect to the receiver means that the calculation of the position of the satellite needs to be taken into account as well, which can only be handled effectively with a computer.

The Global Positioning System, better known simply as GPS, sends several signals that are used to decode the position and distance of the satellite. One signal encodes the satellite's "ephemeris" data, which is used to accurately calculate the satellite's location at any time. Space weather and other effects causes the orbit to change over time so the ephemeris has to be updated periodically. Other signals send out the time as measured by the satellite's onboard atomic clock. By measuring this signal from several satellites, the receiver can re-build an accurate clock signal of its own. Comparing the two produces the distance to the satellite, and several such measurements allows a form of triangulation to be carried out.

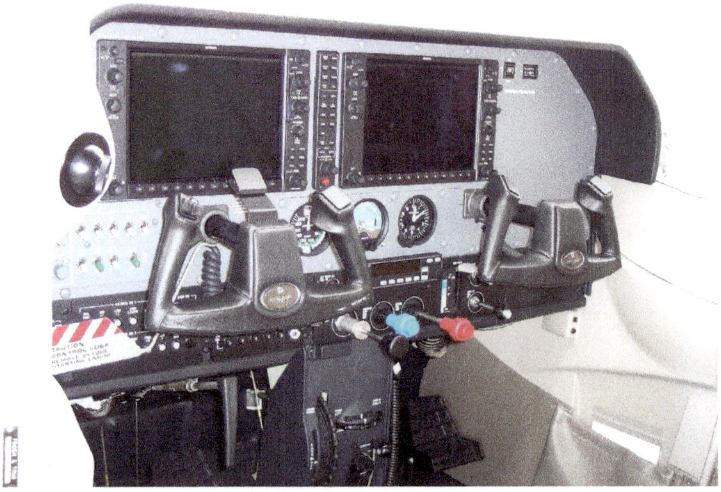

Cessna 182 with GPS-based "glass cockpit" avionics

GPS has better accuracy that any previous land-based system, is available at almost all locations on the Earth, can be implemented in a few cents of modern electronics, and requires only a few dozen satellites to provide worldwide coverage. As a result of these advantages, GPS has led to almost all previous systems falling from use. LORAN, Omega, Decca, Consol and many other systems disappeared during the 1990s and 2000s. The only other systems still in use are aviation aids, which are also being turned off for long-range navigation while new differential GPS systems are being deployed to provide the local accuracy needed for blind landings.

Radar

Long-range radar antenna, used to track space objects and ballistic missiles.

Radar is an object-detection system that uses radio waves to determine the range, angle, or velocity of objects. It can be used to detect aircraft, ships, spacecraft, guided missiles, motor vehicles, weather formations, and terrain. A radar system consists of a transmitter producing electromagnetic waves in the radio or microwaves domain, an emitting antenna, a receiving antenna (separate or the same as the previous one) to capture any returns from objects in the path of the emitted signal, a receiver and processor to determine properties of the object(s).

Radar of the type used for detection of aircraft. It rotates steadily, sweeping the airspace with a narrow beam.

Radar was secretly developed by several nations in the period before and during World War II. The term *RADAR* was coined in 1940 by the United States Navy as an acronym for RAdio Detection And Ranging. The term *radar* has since entered English and other languages as a common noun, losing all capitalization.

The modern uses of radar are highly diverse, including air and terrestrial traffic control, radar astronomy, air-defence systems, antimissile systems, marine radars to locate landmarks and other ships, aircraft anticollision systems, ocean surveillance systems, outer space surveillance and rendezvous systems, meteorological precipitation monitoring, altimetry and flight control systems, guided missile target locating systems, ground-penetrating radar for geological observations, and range-controlled radar for public health surveillance. High tech radar systems are associated with digital signal processing, machine learning and are capable of extracting useful information from very high noise levels.

Other systems similar to radar make use of other parts of the electromagnetic spectrum. One example is “lidar”, which uses ultraviolet, visible, or near infrared light from lasers rather than radio waves.

History

As early as 1886, German physicist Heinrich Hertz showed that radio waves could be reflected from solid objects. In 1895, Alexander Popov, a physics instructor at the Imperial Russian Navy school in Kronstadt, developed an apparatus using a coherer tube for detecting distant lightning strikes. The next year, he added a spark-gap transmitter. In 1897, while testing this equipment for communicating between two ships in the Baltic Sea, he took note of an interference beat caused by the passage of a third vessel. In his report, Popov wrote that this phenomenon might be used for detecting objects, but he did nothing more with this observation.

The German inventor Christian Hülsmeyer was the first to use radio waves to detect "the presence of distant metallic objects". In 1904 he demonstrated the feasibility of detecting a ship in dense fog, but not its distance from the transmitter. He obtained a patent for his detection device in April 1904 and later a patent for a related amendment for estimating the distance to the ship. He also got a British patent on September 23, 1904 for a full radar system, that he called a *telemobiloscope*. It operated on a 50 cm wavelength and the pulsed radar signal was created via a spark-gap. His system already used the classic antenna setup of horn antenna with parabolic reflector and was presented to German military officials in practical tests in Cologne and Rotterdam harbour but was rejected.

Experimental radar antenna, US Naval Research Laboratory, Anacostia, D. C., late 1930s

The first workable radar unit constructed by Robert Watson Watt and his team

In 1922 A. Hoyt Taylor and Leo C. Young, researchers working with the U.S. Navy, had a transmitter and a receiver on opposite sides of the Potomac River and discovered that a ship passing through the beam path caused the received signal to fade in and out. Taylor submitted a report, suggesting that this might be used to detect the presence of ships in low visibility, but the Navy did not immediately continue the work. Eight years later, Lawrence A. Hyland at the Naval Research Laboratory observed similar fading effects from a passing aircraft; this led to a patent application as well as a proposal for serious work at the NRL (Taylor and Young were then at this laboratory) on radio-echo signals from moving targets.

A Chain Home tower in Great Baddow, United Kingdom

During the 1920s the UK research establishment made many advances using radio techniques, including the probing of the ionosphere and the detection of lightning at long distances. Robert Watson-Watt became an expert on the use of radio direction finding as part of his lightning experiments. As part of ongoing experiments, he asked the "new boy", Arnold Frederic Wilkins, to find a receiver suitable for use with shortwave transmissions. Wilkins made an extensive study of available units before selecting a model from the General Post Office. Its instruction manual noted that there was "fading" (the common term for interference at the time) when aircraft flew by.

Before the Second World War, researchers in France, Germany, Italy, Japan, the Netherlands, the Soviet Union, the United Kingdom, and the United States, independently and in great secrecy, developed technologies that led to the modern version of radar. Australia, Canada, New Zealand, and South Africa followed prewar Great Britain, and Hungary had similar developments during the war.

In France in 1934, following systematic studies on the Split Anode Magnetron, the research branch of the Compagnie Générale de Télégraphie Sans Fil (CSF), headed by Maurice Ponte, with Henri Gutton, Sylvain Berline, and M. Hugon, began developing an obstacle-locating radio apparatus, a part of which was installed on the liner *Normandie* in 1935.

During the same time, the Soviet military engineer P. K. Oshchepkov, in collaboration with Leningrad Electrophysical Institute, produced an experimental apparatus, RAPID, capable of detecting an aircraft within 3 km of a receiver. The Soviets produced their first mass production radars RUS-1 and RUS-2 Redut in 1939 but further development was slowed by the NKVD arrest of Oshchepkov and their sending him to the gulag. In total, only 607 Redut stations were produced during the war. The first Russian airborne radar, Gneiss-2, entered into service in June 1943 on Pe-2 fighters. More than 230 Gneiss-2 stations were produced by the end of 1944. The French and Soviet systems, however, had continuous-wave operation and could not give the full performance that was ultimately at the center of modern radar.

Full radar evolved as a pulsed system, and the first such elementary apparatus was demonstrated in December 1934 by the American Robert M. Page, working at the Naval Research Laboratory. The following year, the United States Army successfully tested a primitive surface-to-surface radar to aim coastal battery search lights at night. This was followed by a pulsed system demonstrated in May 1935 by Rudolf Kühnhold and the firm GEMA in Germany and then one in June 1935 by an Air Ministry team led by Robert A. Watson-Watt in Great Britain. Development of radar greatly expanded on 1 September 1936 when Watson-Watt became Superintendent of a new establishment under the British Air Ministry, Bawdsey Research Station located in Bawdsey Manor, near Felixstowe, Suffolk. Work there resulted in the design and installation of aircraft detection and tracking stations called "Chain Home" along the East and South coasts of England in time for the outbreak of World War II in 1939. This system provided the vital advance information that helped the Royal Air Force win the Battle of Britain.

In 1935 Watt was asked to pass judgement on recent reports of a German radio-based death ray and turned the request over to Wilkins. Wilkins returned a set of calculations demonstrating the system was basically impossible. When Watt then asked what might they do, Wilkins recalled the earlier report about aircraft causing radio interference. This led to the Daventry Experiment of 26 February 1935, using a powerful BBC shortwave transmitter as the source and their GPO receiver set up in a field while a bomber flew around the site. When returns were clearly seen, funds were immediately provided for development of an operational system. Watt's team patented the device in GB593017.

Given all required funding and development support, the team had working radar systems in 1935 and began deployment. By 1936 the first five Chain Home (CH) systems were operational and by 1940 stretched across the entire UK including Northern Ireland. Even by standards of the era, CH was crude; instead of broadcasting and receiving from an aimed antenna, CH broadcast a signal floodlighting the entire area in front of it, and then used one of Watt's own radio direction finders to determine the direction of the returned echoes. This meant that CH transmitters had to be much more powerful and have better antennas than competing systems but allowed its rapid introduction using existing technologies.

In April 1940, *Popular Science* showed an example of a radar unit using the Watson-Watt patent in an article on air defence. Also, in late 1941 *Popular Mechanics* had an article in which a U.S. scientist speculated about the British early warning system on the English east coast and came close to what it was and how it worked. Alfred Lee Loomis organized the Radiation Laboratory at Cambridge, Massachusetts which developed the technology in the years 1941-45. Later, in 1943, Page greatly improved radar with the monopulse technique that was used for many years in most radar applications.

The war precipitated research to find better resolution, more portability, and more features for radar, including complementary navigation systems like Oboe used by the RAF's Pathfinder.

Applications

Commercial marine radar antenna. The rotating antenna radiates a vertical fan-shaped beam.

The information provided by radar includes the bearing and range (and therefore position) of the object from the radar scanner. It is thus used in many different fields where the need for such positioning is crucial. The first use of radar was for military purposes: to locate air, ground and sea targets. This evolved in the civilian field into applications for aircraft, ships, and roads.

In aviation, aircraft are equipped with radar devices that warn of aircraft or other obstacles in or approaching their path, display weather information, and give accurate altitude readings. The first commercial device fitted to aircraft was a 1938 Bell Lab unit on some United Air Lines aircraft. Such aircraft can land in fog at airports equipped with radar-assisted ground-controlled approach systems in which the plane's flight is observed on radar screens while operators radio landing directions to the pilot.

Marine radars are used to measure the bearing and distance of ships to prevent collision with other ships, to navigate, and to fix their position at sea when within range of shore or other fixed references such as islands, buoys, and lightships. In port or in harbour, vessel traffic service radar systems are used to monitor and regulate ship movements in busy waters.

Meteorologists use radar to monitor precipitation and wind. It has become the primary tool for short-term weather forecasting and watching for severe weather such as thunderstorms, tornadoes, winter storms, precipitation types, etc. Geologists use specialized ground-penetrating radars to map the composition of Earth's crust. Police forces use radar guns to monitor vehicle speeds on the roads. Smaller radar systems are used to detect human movement. Examples are breathing pattern detection for sleep monitoring and hand and finger gesture detection for computer interaction.

Principles

Radar Signal

A radar system has a transmitter that emits radio waves called *radar signals* in predetermined directions. When these come into contact with an object they are usually reflected or scattered in many directions. Radar signals are reflected especially well by materials of considerable electrical conductivity—especially by most metals, by seawater and by wet ground. Some of these make the use of radar altimeters possible. The radar signals that are reflected back towards the transmitter are the desirable ones that make radar work. If the object is *moving* either toward or away from the transmitter, there is a slight equivalent change in the frequency of the radio waves, caused by the Doppler effect.

Radar receivers are usually, but not always, in the same location as the transmitter. Although the reflected radar signals captured by the receiving antenna are usually very weak, they can be strengthened by electronic amplifiers. More sophisticated methods of signal processing are also used in order to recover useful radar signals.

The weak absorption of radio waves by the medium through which it passes is what enables radar sets to detect objects at relatively long ranges—ranges at which other electromagnetic wavelengths, such as visible light, infrared light, and ultraviolet light, are too strongly attenuated. Such weather phenomena as fog, clouds, rain, falling snow, and sleet that block visible light are usually transparent to radio waves. Certain radio frequencies that are absorbed or scattered by water vapour, raindrops, or atmospheric gases (especially oxygen) are avoided in designing radars, except when their detection is intended.

Illumination

Radar relies on its own transmissions rather than light from the Sun or the Moon, or from electromagnetic waves emitted by the objects themselves, such as infrared wavelengths (heat). This process of directing artificial radio waves towards objects is called *illumination*, although radio waves are invisible to the human eye or optical cameras.

Reflection

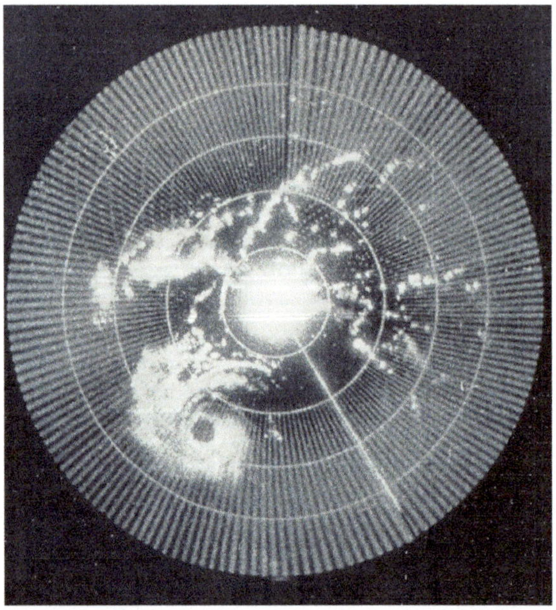

Brightness can indicate reflectivity as in this 1960 weather radar image (of Hurricane Abby). The radar's frequency, pulse form, polarization, signal processing, and antenna determine what it can observe.

If electromagnetic waves travelling through one material meet another material, having a different dielectric constant or diamagnetic constant from the first, the waves will reflect or scatter from the boundary between the materials. This means that a solid object in air or in a vacuum, or a significant change in atomic density between the object and what is surrounding it, will usually scatter radar (radio) waves from its surface. This is particularly true for electrically conductive materials such as metal and carbon fibre, making radar well-suited to the detection of aircraft and ships. Radar absorbing material, containing resistive and sometimes magnetic substances, is used on military vehicles to reduce radar reflection. This is the radio equivalent of painting something a dark colour so that it cannot be seen by the eye at night.

Radar waves scatter in a variety of ways depending on the size (wavelength) of the radio wave and the shape of the target. If the wavelength is much shorter than the target's size, the wave will bounce off in a way similar to the way light is reflected by a mirror. If the wavelength is much longer than the size of the target, the target may not be visible because of poor reflection. Low-frequency radar technology is dependent on resonances for detection, but not identification, of targets. This is described by Rayleigh scattering, an effect that creates Earth's blue sky and red sunsets. When the two length scales are comparable, there may be resonances. Early radars used very long wavelengths that were larger than the targets and thus received a vague signal, where as some modern systems use shorter wavelengths (a few centimetres or less) that can image objects as small as a loaf of bread.

Short radio waves reflect from curves and corners in a way similar to glint from a rounded piece of glass. The most reflective targets for short wavelengths have 90° angles between the reflective surfaces. A corner reflector consists of three flat surfaces meeting like the inside corner of a box. The structure will reflect waves entering its opening directly back to the source. They are commonly used as radar reflectors to make otherwise difficult-to-detect objects easier to detect. Corner reflectors on boats, for example, make them more detectable to avoid collision or during a rescue. For similar reasons, objects intended to avoid detection will not have inside corners or surfaces and edges perpendicular to likely detection directions, which leads to "odd" looking stealth aircraft. These precautions do not completely eliminate reflection because of diffraction, especially at longer wavelengths. Half wavelength long wires or strips of conducting material, such as chaff, are very reflective but do not direct the scattered energy back toward the source. The extent to which an object reflects or scatters radio waves is called its radar cross section.

Radar Equation

The power P_r returning to the receiving antenna is given by the equation:

$$P_r = \frac{P_t G_t A_r \sigma F^4}{(4\pi)^2 R_t^2 R_r^2}$$

where

- P_t = transmitter power
- G_t = gain of the transmitting antenna
- A_r = effective aperture (area) of the receiving antenna; this can also be expressed as $\frac{G_r \lambda^2}{4\pi}$, , where

    - = transmitted wavelength
    - G_r = gain of receiving antenna
- σ = radar cross section, or scattering coefficient, of the target
- F = pattern propagation factor
- R_t = distance from the transmitter to the target
- R_r = distance from the target to the receiver.

In the common case where the transmitter and the receiver are at the same location, $R_t = R_r$ and the term $R_t^2\, R_r^2$ can be replaced by R^4, where R is the range. This yields:

$$P_r = \frac{P_t G_t A_r \sigma F^4}{(4\pi)^2 R^4}.$$

This shows that the received power declines as the fourth power of the range, which means that the received power from distant targets is relatively very small.

Additional filtering and pulse integration modifies the radar equation slightly for pulse-Doppler radar performance, which can be used to increase detection range and reduce transmit power.

The equation above with $F = 1$ is a simplification for transmission in a vacuum without interference. The propagation factor accounts for the effects of multipath and shadowing and depends on the details of the environment. In a real-world situation, pathloss effects should also be considered.

Doppler Effect

Frequency shift is caused by motion that changes the number of wavelengths between the reflector and the radar. That can degrade or enhance radar performance depending upon how that affects the detection process. As an example, Moving Target Indication can interact with Doppler to produce signal cancellation at certain radial velocities, which degrades performance.

Sea-based radar systems, semi-active radar homing, active radar homing, weather radar, military aircraft, and radar astronomy rely on the Doppler effect to enhance performance. This produces information about target velocity during the detection process. This also allows small objects to be detected in an environment containing much larger nearby slow moving objects.

Doppler shift depends upon whether the radar configuration is active or passive. Active radar transmits a signal that is reflected back to the receiver. Passive radar depends upon the object sending a signal to the receiver.

The Doppler frequency shift for active radar is as follows, where F_D is Doppler frequency, F_D is transmit frequency V_R, is radial velocity, and C is the speed of light:

$$F_D = 2 \times F_T \times \left(\frac{V_R}{C}\right).$$

Passive radar is applicable to electronic countermeasures and radio astronomy as follows:

$$F_D = F_T \times \left(\frac{V_R}{C}\right).$$

Only the radial component of the velocity is relevant. When the reflector is moving at right angle to the radar beam, it has no relative velocity. Vehicles and weather moving parallel to the radar beam produce the maximum Doppler frequency shift.

When the transmit frequency (F_T) is pulsed, using a pulse repeat frequency of F_R, the resulting frequency spectrum will contain harmonic frequencies above and below F_T with a distance of F_R. As a result, the Doppler measurement is only non-ambiguous if the Doppler frequency shift is less than half of F_R, called the Nyquist frequency, since the returned frequency otherwise cannot be distinguished from shifting of a harmonic frequency above or below, thus requiring:

$$|F_D| < \frac{F_R}{2}$$

Or when substituting with F_D:

$$|V_R| < \frac{F_R \times \frac{C}{F_T}}{4}$$

As an example, a Doppler weather radar with a pulse rate of 2 kHz and transmit frequency of 1 GHz can reliably measure weather speed up to at most 150 m/s (340 mph), thus cannot reliably determine radial velocity of aircraft moving 1,000 m/s (2,200 mph).

Polarization

In all electromagnetic radiation, the electric field is perpendicular to the direction of propagation, and the electric field direction is the polarization of the wave. For a transmitted radar signal, the polarization can be controlled to yield different effects. Radars use horizontal, vertical, linear, and circular polarization to detect different types of reflections. For example, circular polarization is used to minimize the interference caused by rain. Linear polarization returns usually indicate metal surfaces. Random polarization returns usually indicate a fractal surface, such as rocks or soil, and are used by navigation radars.

Limiting Factors

Beam Path and Range

The radar beam would follow a linear path in vacuum, but it really follows a somewhat curved path in the atmosphere because of the variation of the refractive index of air, that is called the radar horizon. Even when the beam is emitted parallel to the ground, it will rise above it as the Earth curvature sinks below the horizon. Furthermore, the signal is attenuated by the medium it crosses, and the beam disperses.

BEAM HEIGHT WITH DISTANCE (AGL)

$$H = \left(\sqrt{r^2 + (k_e a_e)^2 + 2r k_e a_e \sin(\theta_e)}\right) - k_e a_e + h_a$$

r : distance *k_e : 4/3 (Standard refraction coefficient)*

a_e : Earth radius *θ_e : Elevation angle*

h_a : Height of radar above ground

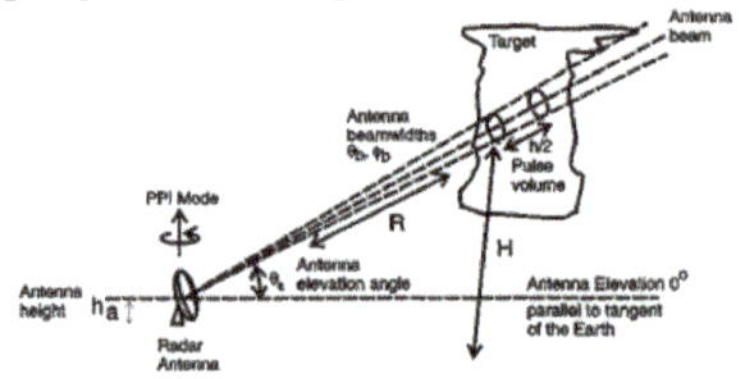

Echo heights above ground

The maximum range of a conventional radar can be limited by a number of factors:

- Line of sight, which depends on height above ground. This means without a direct line of sight the path of the beam is blocked.
- The maximum non-ambiguous range, which is determined by the pulse repetition frequency. The maximum non-ambiguous range is the distance the pulse could travel and return before the next pulse is emitted.
- Radar sensitivity and power of the return signal as computed in the radar equation. This includes factors such as environmental conditions and the size (or radar cross section) of the target.

Noise

Signal noise is an internal source of random variations in the signal, which is generated by all electronic components.

Reflected signals decline rapidly as distance increases, so noise introduces a radar range limitation. The noise floor and signal to noise ratio are two different measures of performance that affect range performance. Reflectors that are too far away produce too little signal to exceed the noise floor and cannot be detected. Detection requires a signal that exceeds the noise floor by at least the signal to noise ratio.

Noise typically appears as random variations superimposed on the desired echo signal received in the radar receiver. The lower the power of the desired signal, the more difficult it is to discern it from the noise. Noise figure is a measure of the noise produced by a receiver compared to an ideal receiver, and this needs to be minimized.

Shot noise is produced by electrons in transit across a discontinuity, which occurs in all detectors. Shot noise is the dominant source in most receivers. There will also be flicker noise caused by electron transit through amplification devices, which is reduced using heterodyne amplification. Another reason for heterodyne processing is that for fixed fractional bandwidth, the instantaneous bandwidth increases linearly in frequency. This allows improved range resolution. The one notable exception to heterodyne (downconversion) radar systems is ultra-wideband radar. Here a single cycle, or transient wave, is used similar to UWB communications.

Noise is also generated by external sources, most importantly the natural thermal radiation of the background surrounding the target of interest. In modern radar systems, the internal noise is typically about equal to or lower than the external noise. An exception is if the radar is aimed upwards at clear sky, where the scene is so "cold" that it generates very little thermal noise. The thermal noise is given by $k_B\ T\ B$, where T is temperature, B is bandwidth (post matched filter) and k_B is Boltzmann's constant. There is an appealing intuitive interpretation of this relationship in a radar. Matched filtering allows the entire energy received from a target to be compressed into a single bin (be it a range, Doppler, elevation, or azimuth bin). On the surface it would appear that then within a fixed interval of time one could obtain perfect, error free, detection. To do this one simply compresses all energy into an infinitesimal time slice. What limits this approach in the real world is that, while time is arbitrarily divisible, current is not. The quantum of electrical energy is an electron, and so the best one can do is match filter all energy into a single electron. Since the electron is moving at a certain temperature (Plank spectrum) this noise source cannot be further eroded. We see then that radar, like all macro-scale entities, is profoundly impacted by quantum theory.

Noise is random and target signals are not. Signal processing can take advantage of this phenomenon to reduce the noise floor using two strategies. The kind of signal integration used with moving target indication can improve noise up to $\sqrt{2}$ for each stage. The signal can also be split among multiple filters for pulse-Doppler signal processing, which reduces the noise floor by the number of filters. These improvements depend upon coherence.

Interference

Radar systems must overcome unwanted signals in order to focus on the targets of interest. These unwanted signals may originate from internal and external sources, both passive and active. The ability of the radar system to overcome these unwanted signals defines its signal-to-noise ratio (SNR). SNR is defined as the ratio of the signal power to the noise power within the desired signal; it compares the level of a desired target signal to the level of background noise (atmospheric noise and noise generated within the receiver). The higher a system's SNR the better it is at discriminating actual targets from noise signals.

Clutter

Clutter refers to radio frequency (RF) echoes returned from targets which are uninteresting to the radar operators. Such targets include natural objects such as ground, sea, precipitation (such as rain, snow or hail), sand storms, animals (especially birds), atmospheric turbulence, and other atmospheric effects, such as ionosphere reflections, meteor trails, and Hail spike. Clutter may also be returned from man-made objects such as buildings and, intentionally, by radar countermeasures such as chaff.

Some clutter may also be caused by a long radar waveguide between the radar transceiver and the antenna. In a typical plan position indicator (PPI) radar with a rotating antenna, this will usually be seen as a "sun" or "sunburst" in the centre of the display as the receiver responds to echoes from dust particles and misguided RF in the waveguide. Adjusting the timing between when the transmitter sends a pulse and when the receiver stage is enabled will generally reduce the sunburst without affecting the accuracy of the range, since most sunburst is caused by a diffused transmit pulse reflected before it leaves the antenna. Clutter is considered a passive interference source, since it only appears in response to radar signals sent by the radar.

Clutter is detected and neutralized in several ways. Clutter tends to appear static between radar scans; on subsequent scan echoes, desirable targets will appear to move, and all stationary echoes can be eliminated. Sea clutter can be reduced by using horizontal polarization, while rain is reduced with circular polarization (note that meteorological radars wish for the opposite effect, and therefore use linear polarization to detect precipitation). Other methods attempt to increase the signal-to-clutter ratio.

Clutter moves with the wind or is stationary. Two common strategies to improve measure or performance in a clutter environment are:

- Moving target indication, which integrates successive pulses and
- Doppler processing, which uses filters to separate clutter from desirable signals.

The most effective clutter reduction technique is pulse-Doppler radar. Doppler separates clutter from aircraft and spacecraft using a frequency spectrum, so individual signals can be separated from multiple reflectors located in the same volume using velocity differences. This requires a coherent transmitter. Another technique uses a moving target indicator that subtracts the receive signal from two successive pulses using phase to reduce signals from slow moving objects. This can be adapted for systems that lack a coherent transmitter, such as time-domain pulse-amplitude radar.

Constant false alarm rate, a form of automatic gain control (AGC), is a method that relies on clutter returns far outnumbering echoes from targets of interest. The receiver's gain is automatically adjusted to maintain a constant level of overall visible clutter. While this does not help detect targets masked by stronger surrounding clutter, it does help to distinguish strong target sources. In the past, radar AGC was electronically controlled and affected the gain of the entire radar receiver. As radars evolved, AGC became computer-software controlled and affected the gain with greater granularity in specific detection cells.

Clutter may also originate from multipath echoes from valid targets caused by ground reflection, atmospheric ducting or ionospheric reflection/refraction (e.g., anomalous propagation). This clutter type is especially bothersome since it appears to move and behave like other normal (point) targets of interest. In a typical scenario, an aircraft echo is reflected from the ground below, appearing to the receiver as an identical target below the correct one. The radar may try to unify the targets, reporting the target at an incorrect height, or eliminating it on the basis of jitter or a physical impossibility. Terrain bounce jamming exploits this response by amplifying the radar signal and directing it downward. These problems can be overcome by incorporating a ground map of the radar's surroundings and eliminating all echoes which appear to originate below ground or above a certain height. Monopulse can be improved by altering the elevation algorithm used at low elevation. In newer air traffic control radar equipment, algorithms are used to identify the false targets by comparing the current pulse returns to those adjacent, as well as calculating return improbabilities.

Jamming

Radar jamming refers to radio frequency signals originating from sources outside the radar, transmitting in the radar's frequency and thereby masking targets of interest. Jamming may be intentional, as with an electronic warfare tactic, or unintentional, as with friendly forces operating equipment that transmits using the same frequency range. Jamming is considered an active interference source, since it is initiated by elements outside the radar and in general unrelated to the radar signals.

Jamming is problematic to radar since the jamming signal only needs to travel one way (from the jammer to the radar receiver) whereas the radar echoes travel two ways (radar-target-radar) and are therefore significantly reduced in power by the time they return to the radar receiver. Jammers therefore can be much less powerful than their jammed radars and still effectively mask targets along the line of sight from the jammer to the radar (*mainlobe jamming*). Jammers have an added effect of affecting radars along other lines of sight through the radar receiver's sidelobes (*sidelobe jamming*).

Mainlobe jamming can generally only be reduced by narrowing the mainlobe solid angle and cannot fully be eliminated when directly facing a jammer which uses the same frequency and polarization as the radar. Sidelobe jamming can be overcome by reducing receiving sidelobes in the radar antenna design and by using an omnidirectional antenna to detect and disregard non-mainlobe signals. Other anti-jamming techniques are frequency hopping and polarization.

Radar Signal Processing

Distance Measurement

Transit Time

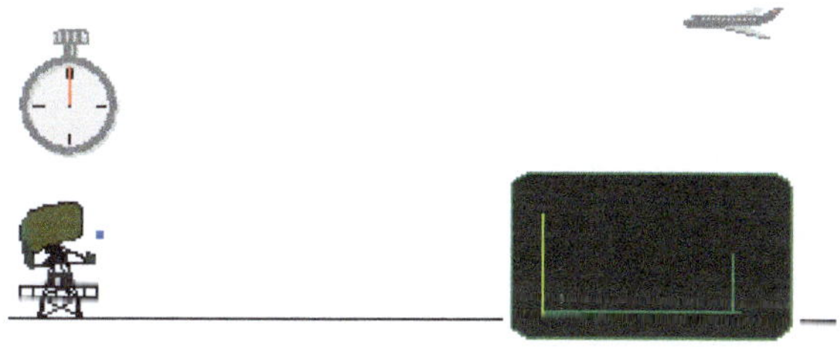

Pulse radar: The round-trip time for the radar pulse to get to the target and return is measured. The distance is proportional to this time.

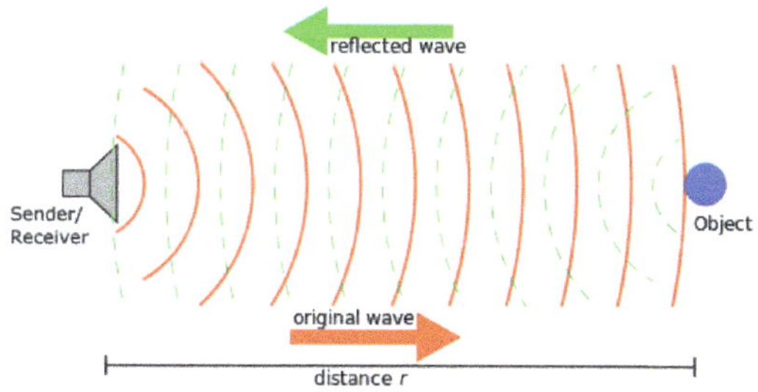

Continuous wave (CW) radar

One way to obtain a distance measurement is based on the time-of-flight: transmit a short pulse of radio signal (electromagnetic radiation) and measure the time it takes for the reflection to return. The distance is one-half the product of the round trip time (because the signal has to travel to the target and then back to the receiver) and the speed of the signal. Since radio waves travel at the speed of light, accurate distance measurement requires high-speed electronics. In most cases, the receiver does not detect the return while the signal is being transmitted. Through the use of a duplexer, the radar switches between transmitting and receiving at a predetermined rate. A similar effect imposes a maximum range as well. In order to maximize range, longer times between pulses should be used, referred to as a pulse repetition time, or its reciprocal, pulse repetition frequency.

These two effects tend to be at odds with each other, and it is not easy to combine both good short range and good long range in a single radar. This is because the short pulses needed for a good minimum range broadcast have less total energy, making the returns much smaller and the target harder to detect. This could be offset by using more pulses, but this would shorten the maximum range. So each radar uses a particular type of signal. Long-range radars tend to use long pulses with long delays between them, and short range radars use smaller pulses with less time between them. As electronics have improved many radars now can change their pulse repetition frequency, thereby changing their range. The newest radars fire two pulses during one cell, one for short range (about 10 km (6.2 mi)) and a separate signal for longer ranges (about 100 km (62 mi)).

The distance resolution and the characteristics of the received signal as compared to noise depends on the shape of the pulse. The pulse is often modulated to achieve better performance using a technique known as pulse compression.

Distance may also be measured as a function of time. The radar mile is the amount of time it takes for a radar pulse to travel one nautical mile, reflect off a target, and return to the radar antenna. Since a nautical mile is defined as 1,852 m, then dividing this distance by the speed of light (299,792,458 m/s), and then multiplying the result by 2 yields a result of 12.36 μs in duration.

Frequency Modulation

Another form of distance measuring radar is based on frequency modulation. Frequency comparison between two signals is considerably more accurate, even with older electronics, than timing the signal. By measuring the frequency of the returned signal and comparing that with the original, the difference can be easily measured.

This technique can be used in continuous wave radar and is often found in aircraft radar altimeters. In these systems a "carrier" radar signal is frequency modulated in a predictable way, typically varying up and down with a sine wave or sawtooth pattern at audio frequencies. The signal is then sent out from one antenna and received on another, typically located on the bottom of the aircraft, and the signal can be continuously compared using a simple *beat frequency* modulator that produces an audio frequency tone from the returned signal and a portion of the transmitted signal.

Since the signal frequency is changing, by the time the signal returns to the aircraft the transmit frequency has changed. The amount of frequency shift is used to measure distance.

The modulation index riding on the receive signal is proportional to the time delay between the radar and the reflector. The amount of that frequency shift becomes greater with greater time delay. The measure of the amount of frequency shift is directly proportional to the distance travelled. That distance can be displayed on an instrument, and it may also be available via the transponder. This signal processing is similar to that used in speed detecting Doppler radar. Example systems using this approach are AZUSA, MISTRAM, and UDOP.

A further advantage is that the radar can operate effectively at relatively low frequencies. This was important in the early development of this type when high frequency signal generation was difficult or expensive.

Terrestrial radar uses low-power FM signals that cover a larger frequency range. The multiple reflections are analyzed mathematically for pattern changes with multiple passes creating a computerized synthetic image. Doppler effects are used which allows slow moving objects to be detected as well as largely eliminating "noise" from the surfaces of bodies of water.

Speed Measurement

Speed is the change in distance to an object with respect to time. Thus the existing system for measuring distance, combined with a memory capacity to see where the target last was, is enough to measure speed. At one time the memory consisted of a user making grease pencil marks on the radar screen and then calculating the speed using a slide rule. Modern radar systems perform the equivalent operation faster and more accurately using computers.

If the transmitter's output is coherent (phase synchronized), there is another effect that can be used to make almost instant speed measurements (no memory is required), known as the Doppler effect. Most modern radar systems use this principle into Doppler radar and pulse-Doppler radar systems (weather radar, military radar). The Doppler effect is only able to determine the relative speed of the target along the line of sight from the radar to the target. Any component of target velocity perpendicular to the line of sight cannot be determined by using the Doppler effect alone, but it can be determined by tracking the target's azimuth over time.

It is possible to make a Doppler radar without any pulsing, known as a continuous-wave radar (CW radar), by sending out a very pure signal of a known frequency. CW radar is ideal for determining the radial component of a target's velocity. CW radar is typically used by traffic enforcement to measure vehicle speed quickly and accurately where range is not important.

When using a pulsed radar, the variation between the phase of successive returns gives the distance the target has moved between pulses, and thus its speed can be calculated. Other mathematical developments in radar signal processing include time-frequency analysis (Weyl Heisenberg or wavelet), as well as the chirplet transform which makes use of the change of frequency of returns from moving targets ("chirp").

Pulse-Doppler Signal Processing

Pulse-Doppler signal processing includes frequency filtering in the detection process. The space between each transmit pulse is divided into range cells or range gates. Each cell is filtered independently much like the process used by a spectrum analyzer to produce the display showing different frequencies. Each different distance produces a different spectrum. These spectra are used to perform the detection process. This is required to achieve acceptable performance in hostile environments involving weather, terrain, and electronic countermeasures.

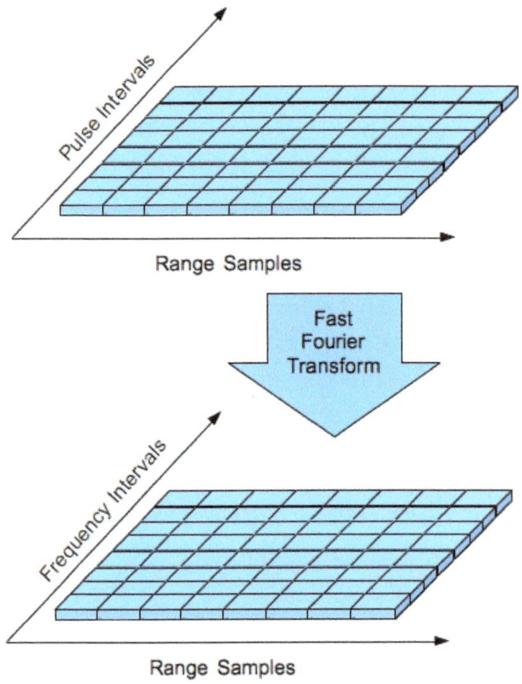

Pulse-Doppler signal processing. The *Range Sample* axis represents individual samples taken in between each transmit pulse. The *Range Interval* axis represents each successive transmit pulse interval during which samples are taken. The Fast Fourier Transform process converts time-domain samples into frequency domain spectra. This is sometimes called the *bed of nails*.

The primary purpose is to measure both the amplitude and frequency of the aggregate reflected signal from multiple distances. This is used with weather radar to measure radial wind velocity and precipitation rate in each different volume of air. This is linked with computing systems to produce a real-time electronic weather map. Aircraft safety depends upon continuous access to accurate weather radar information that is used to prevent injuries and accidents. Weather radar uses a low PRF. Coherency requirements are not as strict as those for military systems because individual signals ordinarily do not need to be separated. Less sophisticated filtering is required, and range ambiguity processing is not normally needed with weather radar in comparison with military radar intended to track air vehicles.

The alternate purpose is "look-down/shoot-down" capability required to improve military air combat survivability. Pulse-Doppler is also used for ground based surveillance radar required to defend personnel and vehicles. Pulse-Doppler signal processing increases the maximum detection distance using less radiation in close proximity to aircraft pilots, shipboard personnel, infantry, and artillery. Reflections from terrain, water, and weather produce signals much larger than aircraft and missiles, which allows fast moving vehicles to hide using nap-of-the-earth flying techniques and stealth technology to avoid detection until an attack vehicle is too close to destroy. Pulse-Doppler signal processing incorporates more sophisticated electronic filtering that safely eliminates this kind of weakness. This requires the use of medium pulse-repetition frequency with phase coherent hardware that has a large dynamic range. Military applications require medium PRF which prevents range from being determined directly, and range ambiguity resolution processing is required to identify the true range of all reflected signals. Radial movement is usually linked with Doppler frequency to produce a lock signal that cannot be produced by radar jamming signals. Pulse-Doppler signal processing also produces audible signals that can be used for threat identification.

Reduction of Interference Effects

Signal processing is employed in radar systems to reduce the radar interference effects. Signal processing techniques include moving target indication, Pulse-Doppler signal processing, moving target detection processors, correlation with secondary surveillance radar targets, space-time adaptive processing, and track-before-detect. Constant false alarm rate and digital terrain model processing are also used in clutter environments.

Plot and Track Extraction

A Track algorithm is a radar performance enhancement strategy. Tracking algorithms provide the ability to predict future position of multiple moving objects based on the history of the individual positions being reported by sensor systems.

Historical information is accumulated and used to predict future position for use with air traffic control, threat estimation, combat system doctrine, gun aiming, and missile guidance. Position data is accumulated radar sensors over the span of a few minutes.

There are four common track algorithms.

- Nearest Neighbour
- Probabilistic Data Association
- Multiple Hypothesis Tracking
- Interactive Multiple Model (IMM)

Radar video returns from aircraft can be subjected to a plot extraction process whereby spurious and interfering signals are discarded. A sequence of target returns can be monitored through a device known as a plot extractor.

The non-relevant real time returns can be removed from the displayed information and a single plot displayed. In some radar systems, or alternatively in the command and control system to which the radar is connected, a radar tracker is used to associate the sequence of plots belonging to individual targets and estimate the targets' headings and speeds.

Engineering

A radar's components are:

- A transmitter that generates the radio signal with an oscillator such as a klystron or a magnetron and controls its duration by a modulator.

- A waveguide that links the transmitter and the antenna.
- A duplexer that serves as a switch between the antenna and the transmitter or the receiver for the signal when the antenna is used in both situations.
- A receiver. Knowing the shape of the desired received signal (a pulse), an optimal receiver can be designed using a matched filter.
- A display processor to produce signals for human readable output devices.
- An electronic section that controls all those devices and the antenna to perform the radar scan ordered by software.
- A link to end user devices and displays.

Components of a Radar/Composantes d'un radar

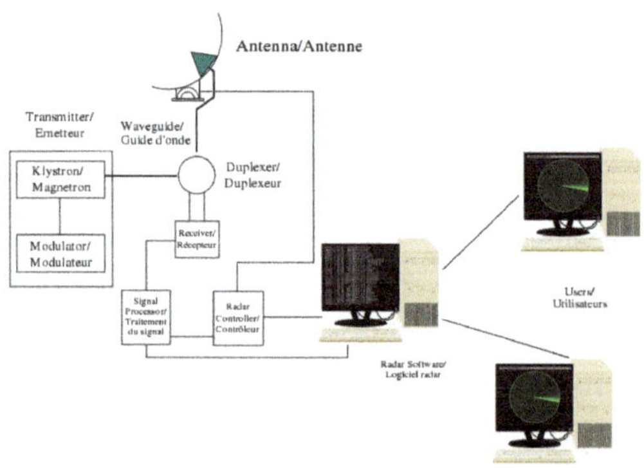

Radar components

Antenna Design

Radio signals broadcast from a single antenna will spread out in all directions, and likewise a single antenna will receive signals equally from all directions. This leaves the radar with the problem of deciding where the target object is located.

Early systems tended to use omnidirectional broadcast antennas, with directional receiver antennas which were pointed in various directions. For instance, the first system to be deployed, Chain Home, used two straight antennas at right angles for reception, each on a different display. The maximum return would be detected with an antenna at right angles to the target, and a minimum with the antenna pointed directly at it (end on). The operator could determine the direction to a target by rotating the antenna so one display showed a maximum while the other showed a minimum. One serious limitation with this type of solution is that the broadcast is sent out in all directions, so the amount of energy in the direction being examined is a small part of that transmitted. To get a reasonable amount of power on the "target", the transmitting aerial should also be directional.

Parabolic Reflector

More modern systems use a steerable parabolic "dish" to create a tight broadcast beam, typically using the same dish as the receiver. Such systems often combine two radar frequencies in the same antenna in order to allow automatic steering, or *radar lock*.

Parabolic reflectors can be either symmetric parabolas or spoiled parabolas: Symmetric parabolic antennas produce a narrow "pencil" beam in both the X and Y dimensions and consequently have a higher gain. The NEXRAD Pulse-Doppler weather radar uses a symmetric antenna to perform detailed volumetric scans of the atmosphere. Spoiled parabolic antennas produce a narrow beam in one dimension and a relatively wide beam in the other. This feature is useful if target detection over a wide range of angles is more important than target location in three dimensions. Most 2D surveillance radars use a spoiled parabolic antenna with a narrow azimuthal beamwidth and wide vertical beamwidth. This beam configuration allows the radar operator to detect an aircraft at a specific azimuth but at an indeterminate height. Conversely, so-called "nodder" height finding radars use a dish with a narrow vertical beamwidth and wide azimuthal beamwidth to detect an aircraft at a specific height but with low azimuthal precision.

Types of scan

- Primary Scan: A scanning technique where the main antenna aerial is moved to produce a scanning beam, examples include circular scan, sector scan, etc.
- Secondary Scan: A scanning technique where the antenna feed is moved to produce a scanning beam, examples include conical scan, unidirectional sector scan, lobe switching, etc.
- Palmer Scan: A scanning technique that produces a scanning beam by moving the main antenna and its feed. A Palmer Scan is a combination of a Primary Scan and a Secondary Scan.
- Conical scanning: The radar beam is rotated in a small circle around the "boresight" axis, which is pointed at the target.

Slotted Waveguide

Slotted waveguide antenna

Applied similarly to the parabolic reflector, the slotted waveguide is moved mechanically to scan and is particularly suitable for non-tracking surface scan systems, where the vertical pattern may remain constant. Owing to its lower cost and less wind exposure, shipboard, airport surface, and harbour surveillance radars now use this approach in preference to a parabolic antenna.

Phased Array

Phased array: Not all radar antennas must rotate to scan the sky.

Another method of steering is used in a phased array radar.

Phased array antennas are composed of evenly spaced similar antenna elements, such as aerials or rows of slotted waveguide. Each antenna element or group of antenna elements incorporates a discrete phase shift that produces a phase gradient across the array. For example, array elements producing a 5 degree phase shift for each wavelength across the array face will produce a beam pointed 5 degree away from the centreline perpendicular to the array face. Signals travelling along that beam will be reinforced. Signals offset from that beam will be cancelled. The amount of reinforcement is antenna gain. The amount of cancellation is side-lobe suppression.

Phased array radars have been in use since the earliest years of radar in World War II (Mammut radar), but electronic device limitations led to poor performance. Phased array radars were originally used for missile defence. They are the heart of the ship-borne Aegis Combat System and the Patriot Missile System. The massive redundancy associated with having a large number of array elements increases reliability at the expense of gradual performance degradation that occurs as individual phase elements fail.

Phased array antenna can be built to conform to specific shapes, like missiles, infantry support vehicles, ships, and aircraft.

As the price of electronics has fallen, phased array radars have become more common. Almost all modern military radar systems are based on phased arrays, where the small additional cost is offset by the improved reliability of a system with no moving parts. Traditional moving-antenna designs are still widely used in roles where cost is a significant factor such as air traffic surveillance, weather radars and similar systems.

Phased array radars are valued for use in aircraft since they can track multiple targets. The first aircraft to use a phased array radar was the B-1B Lancer. The first fighter aircraft to use phased array radar was the Mikoyan MiG-31. The MiG-31M's SBI-16 Zaslon Passive electronically scanned array radar was considered to be the world's most powerful fighter radar, until the AN/APG-77 Active electronically scanned array was introduced on the Lockheed Martin F-22 Raptor.

Phased-array interferometry or aperture synthesis techniques, using an array of separate dishes that are phased into a single effective aperture, are not typical for radar applications, although they are widely used in radio astronomy. Because of the thinned array curse, such multiple aperture arrays, when used in transmitters, result in narrow beams at the expense of reducing the total power transmitted to the target. In principle, such techniques could increase spatial resolution, but the lower power means that this is generally not effective.

Aperture synthesis by post-processing motion data from a single moving source, on the other hand, is widely used in space and airborne radar systems .

Frequency Bands

The traditional band names originated as code-names during World War II and are still in military and aviation use throughout the world. They have been adopted in the United States by the Institute of Electrical and Electronics Engineers and internationally by the International Telecommunication Union. Most countries have additional regulations to control which parts of each band are available for civilian or military use.

Other users of the radio spectrum, such as the broadcasting and electronic countermeasures industries, have replaced the traditional military designations with their own systems.

Radar frequency bands			
Band name	**Frequency range**	**Wavelength range**	**Notes**
HF	3–30 MHz	10–100 m	Coastal radar systems, over-the-horizon radar (OTH) radars; 'high frequency'
VHF	30–300 MHz	1–10 m	Very long range, ground penetrating; 'very high frequency'
P	< 300 MHz	> 1 m	'P' for 'previous', applied retrospectively to early radar systems; essentially HF + VHF
UHF	300–1000 MHz	0.3–1 m	Very long range (e.g. ballistic missile early warning), ground penetrating, foliage penetrating; 'ultra high frequency'
L	1–2 GHz	15–30 cm	Long range air traffic control and surveillance; 'L' for 'long'
S	2–4 GHz	7.5–15 cm	Moderate range surveillance, Terminal air traffic control, long-range weather, marine radar; 'S' for 'short'
C	4–8 GHz	3.75–7.5 cm	Satellite transponders; a compromise (hence 'C') between X and S bands; weather; long range tracking
X	8–12 GHz	2.5–3.75 cm	Missile guidance, marine radar, weather, medium-resolution mapping and ground surveillance; in the United States the narrow range 10.525 GHz ±25 MHz is used for airport radar; short range tracking. Named X band because the frequency was a secret during WW2.

K_u	12–18 GHz	1.67–2.5 cm	High-resolution, also used for satellite transponders, frequency under K band (hence 'u')
K	18–24 GHz	1.11–1.67 cm	From German *kurz*, meaning 'short'; limited use due to absorption by water vapour, so K_u and K_a were used instead for surveillance. K-band is used for detecting clouds by meteorologists, and by police for detecting speeding motorists. K-band radar guns operate at 24.150 ± 0.100 GHz.
K_a	24–40 GHz	0.75–1.11 cm	Mapping, short range, airport surveillance; frequency just above K band (hence 'a') Photo radar, used to trigger cameras which take pictures of license plates of cars running red lights, operates at 34.300 ± 0.100 GHz.
mm	40–300 GHz	1.0–7.5 mm	Millimetre band, subdivided as below. The frequency ranges depend on waveguide size. Multiple letters are assigned to these bands by different groups. These are from Baytron, a now defunct company that made test equipment.
V	40–75 GHz	4.0–7.5 mm	Very strongly absorbed by atmospheric oxygen, which resonates at 60 GHz.
W	75–110 GHz	2.7–4.0 mm	Used as a visual sensor for experimental autonomous vehicles, high-resolution meteorological observation, and imaging.

Radar Modulators

Modulators act to provide the waveform of the RF-pulse. There are two different radar modulator designs:

- High voltage switch for non-coherent keyed power-oscillators These modulators consist of a high voltage pulse generator formed from a high voltage supply, a pulse forming network, and a high voltage switch such as a thyratron. They generate short pulses of power to feed, e.g., the magnetron, a special type of vacuum tube that converts DC (usually pulsed) into microwaves. This technology is known as pulsed power. In this way, the transmitted pulse of RF radiation is kept to a defined and usually very short duration.

- Hybrid mixers, fed by a waveform generator and an exciter for a complex but coherent waveform. This waveform can be generated by low power/low-voltage input signals. In this case the radar transmitter must be a power-amplifier, e.g., a klystron tube or a solid state transmitter. In this way, the transmitted pulse is intrapulse-modulated and the radar receiver must use pulse compression techniques.

Radar Coolant

Coherent microwave amplifiers operating above 1,000 watts microwave output, like travelling wave tubes and klystrons, require liquid coolant. The electron beam must contain 5 to 10 times more power than the microwave output, which can produce enough heat to generate plasma. This plasma flows from the collector toward the cathode. The same magnetic focusing that guides the electron beam forces the plasma into the path of the electron beam but flowing in the opposite direction. This introduces FM modulation which degrades Doppler performance. To prevent this, liquid coolant with minimum pressure and flow rate is required, and deionized water is normally used in most high power surface radar systems that utilize Doppler processing.

Coolanol (silicate ester) was used in several military radars in the 1970s. However, it is hygroscopic, leading to hydrolysis and formation of highly flammable alcohol. The loss of a U.S. Navy aircraft in 1978 was attributed to a silicate ester fire. Coolanol is also expensive and toxic. The U.S. Navy has instituted a program named Pollution Prevention (P2) to eliminate or reduce the volume and toxicity of waste, air emissions, and effluent discharges. Because of this, Coolanol is used less often today.

Regulations

Radar (also: *RADAR*) is defined by *article 1.100* of the International Telecommunication Union´s (ITU) ITU Radio Regulations (RR) as:

A radiodetermination system based on the comparison of reference signals with radio signals reflected, or retransmitted, from the position to be determined. Each *radiodetermination system* shall be classified by the *radiocommunication service* in which it operates permanently or temporarily. Typical radar utilizations are primary radar and secondary radar, these might operate in the radiolocation service or the radiolocation-satellite service.

Wireless Power Transfer

Inductive charging pad for LG smartphone, using the Qi system, an example of near-field wireless transfer. When the phone is set on the pad, a coil in the pad creates a magnetic field which induces a current in another coil, in the phone, charging its battery.

Wireless power transfer (WPT) or wireless energy transmission is the transmission of electrical energy from a power source to an electrical load, such as an electrical power grid or a consuming device, without the use of discrete human-made conductors. Wireless power is a generic term that refers to a number of different power transmission technologies that use time-varying electric, magnetic, or electromagnetic fields. In wireless power transfer, a wireless transmitter connected to a power source conveys the field energy across an intervening space to one or more receivers, where it is converted back to an electrical current and then used. Wireless transmission is useful to power electrical devices in cases where interconnecting wires are inconvenient, hazardous, or are not possible.

Wireless power techniques fall into two categories, non-radiative and radiative. In *non-radiative* techniques, power is typically transferred by magnetic fields using inductive coupling between coils of wire. Applications of this type include electric toothbrush chargers, RFID tags, smartcards, and chargers for implantable medical devices like artificial cardiac pacemakers, and inductive powering or charging of electric vehicles like trains or buses. A current focus is to develop wireless systems to charge mobile and handheld computing devices such as cellphones, digital music players and portable computers without being tethered to a wall plug. Power may also be transferred by electric fields using capacitive coupling between metal electrodes. In *radiative far-field* techniques, also called *power beaming*, power is transferred by beams of electromagnetic radiation, like microwaves or laser beams. These techniques can transport energy longer distances but must be aimed at the receiver. Proposed applications for this type are solar power satellites, and wireless powered drone aircraft.

Japan and China both have national ambitions to begin on-orbit testing of Solar Power Satellites by the 2030s which may accelerate both technical and regulatory progress.

An important issue associated with all wireless power systems is limiting the exposure of people and other living things to potentially injurious electromagnetic fields.

Overview

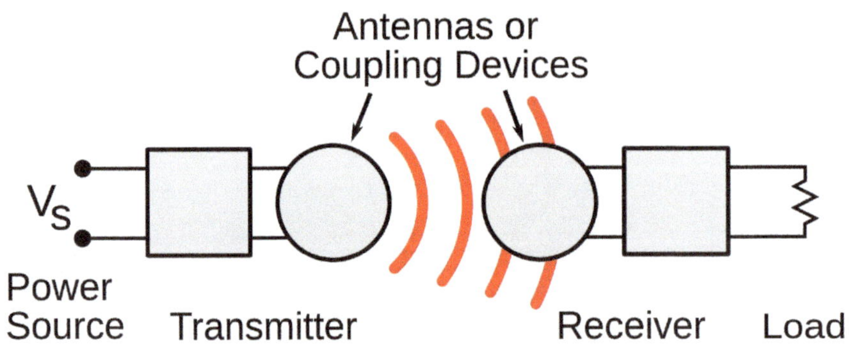

Generic block diagram of a wireless power system

"Wireless power transfer" is a collective term that refers to a number of different technologies for transmitting energy by means of electromagnetic fields. The technologies, listed in the table below, differ in the distance over which they can transfer power efficiently, whether the transmitter must be aimed (directed) at the receiver, and in the type of electromagnetic energy they use: time varying electric fields, magnetic fields, radio waves, microwaves, or infrared or visible light waves.

In general a wireless power system consists of a "transmitter" connected to a source of power such as a mains power line, which converts the power to a time-varying electromagnetic field, and one or more "receiver" devices which receive the power and convert it back to DC or AC electric current which is used by an electrical load. At the transmitter the input power is converted to an oscillating electromagnetic field by some type of "antenna" device. The word "antenna" is used loosely here; it may be a coil of wire which generates a magnetic field, a metal plate which generates an electric field, an antenna which radiates radio waves, or a laser which generates light. A similar antenna or coupling device at the receiver converts the oscillating fields to an electric current. An important parameter that determines the type of waves is the frequency f in hertz of the oscillations. The frequency determines the wavelength $\lambda = c/f$ of the waves which carry the energy across the gap, where **c** is the velocity of light.

Wireless power uses the same fields and waves as wireless communication devices like radio, another familiar technology that involves electrical energy transmitted without wires by electromagnetic fields, used in cellphones, radio and television broadcasting, and WiFi. In radio communication the goal is the transmission of information, so the amount of power reaching the receiver is not so important, as long as it is sufficient so the signal to noise ratio is high enough that the information can be received intelligibly. In wireless communication technologies, generally, only tiny amounts of power reach the receiver. In contrast, with wireless power the amount of energy received is the important thing, so the efficiency (fraction of transmitted energy that is received) is the more significant parameter. For this reason, wireless power technologies are likely to be more limited by distance than wireless communication technologies.

These are the different wireless power technologies:

Technology	Range	Directivity	Frequency	Antenna devices	Current and/or possible future applications
Inductive coupling	Short	Low	Hz – MHz	Wire coils	Electric tooth brush and razor battery charging, induction stovetops and industrial heaters.
Resonant inductive coupling	Mid-	Low	kHz – GHz	Tuned wire coils, lumped element resonators	Charging portable devices (Qi), biomedical implants, electric vehicles, powering busses, trains, MAGLEV, RFID, smartcards.
Capacitive coupling	Short	Low	kHz – MHz	Electrodes	Charging portable devices, power routing in large-scale integrated circuits, Smartcards.
Magnetodynamic coupling	Short	N.A.	Hz	Rotating magnets	Charging electric vehicles, busses, biomedical implants.
Microwaves	Long	High	GHz	Parabolic dishes, phased arrays, rectennas	Solar power satellite, powering drone aircraft.
Light waves	Long	High	≥THz	Lasers, photocells, lenses	Powering drone aircraft, powering space elevator climbers.

Field Regions

Electric and magnetic fields are created by charged particles in matter such as electrons. A stationary charge creates an electrostatic field in the space around it. A steady current of charges (direct current, DC) creates a static magnetic field around it. The above fields contain energy, but cannot carry power because they are static. However time-varying fields can carry power. Accelerating electric charges, such as are found in an alternating current (AC) of electrons in a wire, create time-varying electric and magnetic fields in the space around them. These fields can exert oscillating forces on the electrons in a receiving "antenna", causing them to move back and forth. These represent alternating current which can be used to power a load.

The oscillating electric and magnetic fields surrounding moving electric charges in an antenna device can be divided into two regions, depending on distance $D_{\mathbf{range}}$ from the antenna. The boundary between the regions is somewhat vaguely defined. The fields have different characteristics in these regions, and different technologies are used for transferring power:

- *Near-field* or *nonradiative* region – This means the area within about 1 wavelength (λ) of the antenna. In this region the oscillating electric and magnetic fields are separate and power can be transferred via electric fields by capacitive coupling (electrostatic induction) between metal electrodes, or via magnetic fields by inductive coupling (electromagnetic induction) between coils of wire. These fields are not *radiative*, meaning the energy stays within a short distance of the transmitter. If there is no receiving device or absorbing material within their limited range to "couple" to, no power leaves the transmitter. The range of these fields is short, and depends on the size and shape of the "antenna" devices, which are usually coils of wire. The fields, and thus the power transmitted, decrease exponentially with distance, so if the distance between the two "antennas" $D_{\mathbf{range}}$ is much larger than the diameter of the "antennas" $D_{\mathbf{ant}}$ very little power will be received. Therefore, these techniques cannot be used for long range power transmission.

  Resonance, such as resonant inductive coupling, can increase the coupling between the antennas greatly, allowing efficient transmission at somewhat greater distances, although the fields still decrease exponentially. Therefore the range of near-field devices is conventionally divided into two categories:

  - *Short range* – up to about one antenna diameter: $D_{\mathbf{range}} \leq D_{\mathbf{ant}}$. This is the range over which ordinary nonresonant capacitive or inductive coupling can transfer practical amounts of power.

  - *Mid-range* – up to 10 times the antenna diameter: $D_{\mathbf{range}} \leq 10\, D_{\mathbf{ant}}$. This is the range over which resonant capacitive or inductive coupling can transfer practical amounts of power.

- *Far-field* or *radiative* region – Beyond about 1 wavelength (λ) of the antenna, the electric and magnetic fields are perpendicular to each other and propagate as an electromagnetic wave; examples are radio waves, microwaves, or light waves. This part of the energy is *radiative*, meaning it leaves the antenna whether or not there is a receiver to absorb it. The portion of energy which does not strike the receiving antenna is dissipated and lost to the system. The amount of power emitted as electromagnetic waves by an antenna depends on the ratio of the antenna's size D_{ant} to the wavelength of the waves λ, which is determined by the frequency: $\lambda = c/f$. At low frequencies f where the antenna is much smaller than the size of the waves, $D_{ant} << \lambda$, very little power is radiated. Therefore the near-field devices above, which use lower frequencies, radiate almost none of their energy as electromagnetic radiation. Antennas about the same size as the wavelength $D_{ant} \approx \lambda$ such as monopole or dipole antennas, radiate power efficiently, but the electromagnetic waves are radiated in all directions (omnidirectionally), so if the receiving antenna is far away, only a small amount of the radiation will hit it. Therefore, these can be used for short range, inefficient power transmission but not for long range transmission.

However, unlike fields, electromagnetic radiation can be focused by reflection or refraction into beams. By using a high-gain antenna or optical system which concentrates the radiation into a narrow beam aimed at the receiver, it can be used for *long range* power transmission. From the Rayleigh criterion, to produce the narrow beams necessary to focus a significant amount of the energy on a distant receiver, an antenna must be much larger than the wavelength of the waves used: $D_{ant} >> \lambda = c/f$. Practical *beam power* devices require wavelengths in the centimeter region or below, corresponding to frequencies above 1 GHz, in the microwave range or above.

Non-radiative Techniques

Near-field

At large relative distance, the near-field components of electric and magnetic fields are approximately quasi-static oscillating dipole fields. These fields decrease with the cube of distance: $(D_{range}/D_{ant})^{-3}$ Since power is proportional to the square of the field strength, the power transferred decreases as $(D_{range}/D_{ant})^{-6}$. or 60 dB per decade. In other words, if far apart, doubling the distance between the two antennas causes the power received to decrease by a factor of $2^6 = 64$. As a result, inductive and capacitive coupling can only be used for short-range power transfer, within a few times the diameter of the antenna device D_{ant}. Unlike in a radiative system where the maximum radiation occurs when the dipole antennas are oriented transverse to the direction of propagation, with dipole fields the maximum coupling occurs when the dipoles are oriented longitudinally.

Inductive (Magnetic) Coupling

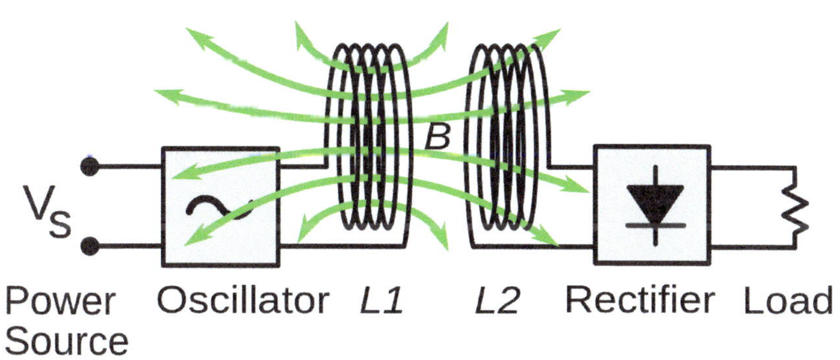

Generic block diagram of an inductive wireless power system.

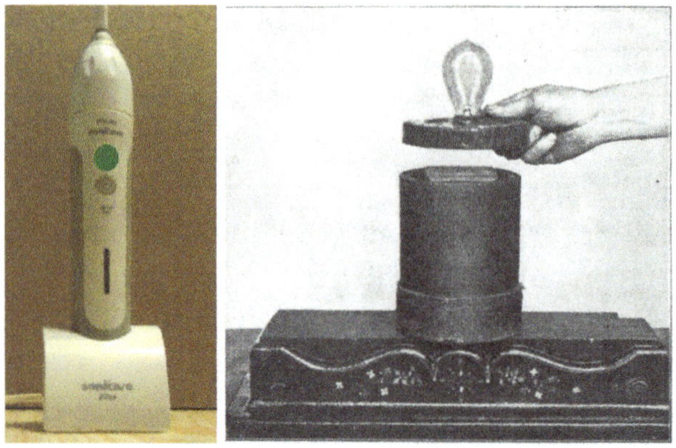

(left) Modern inductive power transfer, an electric toothbrush charger. A coil in the stand produces a magnetic field, inducing an AC current in a coil in the toothbrush, which is rectified to charge the batteries.
(right) A light bulb powered wirelessly by induction, in 1910.

In inductive coupling (*electromagnetic induction* or *inductive power transfer*, IPT), power is transferred between coils of wire by a magnetic field. The transmitter and receiver coils together form a *transformer (see diagram)*. An alternating current (AC) through the transmitter coil *(L1)* creates an oscillating magnetic field *(B)* by Ampere's law. The magnetic field passes through the receiving coil *(L2)*, where it induces an alternating EMF (voltage) by Faraday's law of induction, which creates an AC current in the receiver. The induced alternating current may either drive the load directly, or be rectified to direct current (DC) by a rectifier in the receiver, which drives the load. A few systems, such as electric toothbrush charging stands, work at 50/60 Hz so AC mains current is applied directly to the transmitter coil, but in most systems an electronic oscillator generates a higher frequency AC current which drives the coil, because transmission efficiency improves with frequency.

Inductive coupling is the oldest and most widely used wireless power technology, and virtually the only one so far which is used in commercial products. It is used in inductive charging stands for cordless appliances used in wet environments such as electric toothbrushes and shavers, to reduce the risk of electric shock. Another application area is "transcutaneous" recharging of biomedical prosthetic devices implanted in the human body, such as cardiac pacemakers and insulin pumps, to avoid having wires passing through the skin. It is also used to charge electric vehicles such as cars and to either charge or power transit vehicles like buses and trains.

However the fastest growing use is wireless charging pads to recharge mobile and handheld wireless devices such as laptop and tablet computers, cellphones, digital media players, and video game controllers.

The power transferred increases with frequency and the mutual inductance *M* between the coils, which depends on their geometry and the distance $D_{\mathbf{range}}$ between them. A widely used figure of merit is the coupling coefficient $k = M / \sqrt{L_1 L_2}$. This dimensionless parameter is equal to the fraction of magnetic flux through *L1* that passes through *L2*. If the two coils are on the same axis and close together so all the magnetic flux from *L1* passes through *L2*, $k = 1$ and the link efficiency approaches 100%. The greater the separation between the coils, the more of the magnetic field from the first coil misses the second, and the lower *k* and the link efficiency are, approaching zero at large separations. The link efficiency and power transferred is roughly proportional to k^2. In order to achieve high efficiency, the coils must be very close together, a fraction of the coil diameter D_{ant}, usually within centimeters, with the coils' axes aligned. Wide, flat coil shapes are usually used, to increase coupling. Ferrite "flux confinement" cores can confine the magnetic fields, improving coupling and reducing interference to nearby electronics, but they are heavy and bulky so small wireless devices often use air-core coils.

Ordinary inductive coupling can only achieve high efficiency when the coils are very close together, usually adjacent. In most modern inductive systems resonant inductive coupling *(described below)* is used, in which the efficiency is increased by using resonant circuits. This can achieve high efficiencies at greater distances than nonresonant inductive coupling.

Prototype inductive electric car charging system at 2011 Tokyo Auto Show

Powermat inductive charging spots in a coffee shop. Customers can set their phones and computers on them to recharge.

Resonant Inductive Coupling

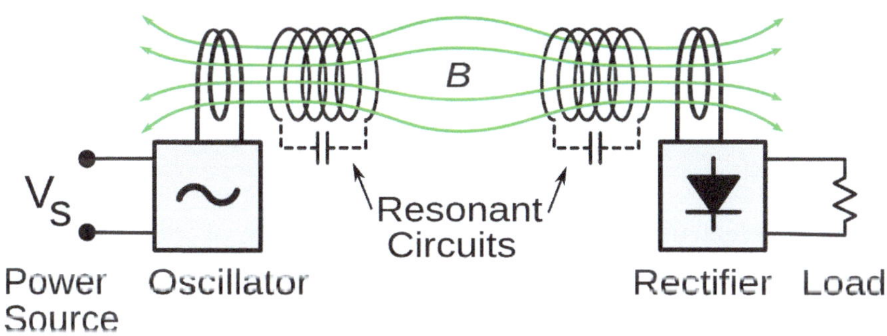

Diagram of the resonant inductive wireless power system demonstrated by Marin Soljačić's MIT team in 2007. The resonant circuits were coils of copper wire which resonated with their internal capacitance (dotted capacitors) at 10 MHz. Power was coupled into the transmitter resonator, and out of the receiver resonator into the rectifier, by small coils which also served for impedance matching.

Resonant inductive coupling (*electrodynamic coupling, strongly coupled magnetic resonance*) is a form of inductive coupling in which power is transferred by magnetic fields *(B, green)* between two resonant circuits (tuned circuits), one in the transmitter and one in the receiver *(see diagram, right)*. Each resonant circuit consists of a coil of wire connected to a capacitor, or a self-resonant coil or other resonator with internal capacitance. The two are tuned to resonate at the same resonant frequency. The resonance between the coils can greatly increase coupling and power transfer, analogously to the way a vibrating tuning fork can induce sympathetic vibration in a distant fork tuned to the same pitch. Nikola Tesla first discovered resonant coupling during his pioneering experiments in wireless power transfer around the turn of the 20th century, but the possibilities of using resonant coupling to increase transmission range has only recently been explored. In 2007 a team led by Marin Soljačić at MIT used two coupled tuned circuits each made of a 25 cm self-resonant coil of wire at 10 MHz to achieve the transmission of 60 W of power over a distance of 2 meters (6.6 ft) (8 times the coil diameter) at around 40% efficiency.

The concept behind resonant inductive coupling is that high Q factor resonators exchange energy at a much higher rate than they lose energy due to internal damping. Therefore, by using resonance, the same amount of power can be transferred at greater distances, using the much weaker magnetic fields out in the peripheral regions ("tails") of the near fields (these are sometimes called evanescent fields). Resonant inductive coupling can achieve high efficiency at ranges of 4 to 10 times the coil diameter (D_{ant}). This is called "mid-range" transfer, in contrast to the "short range" of nonresonant inductive transfer, which can achieve similar efficiencies only when the coils are adjacent. Another advantage is that resonant circuits interact with each other so much more strongly than they do with nonresonant objects that power losses due to absorption in stray nearby objects are negligible. A drawback of resonant coupling is that at close ranges when the two resonant circuits are tightly coupled, the resonant frequency of the system is no longer constant but "splits" into two resonant peaks, so the maximum power transfer no longer occurs at the original resonant frequency and the oscillator frequency must be tuned to the new resonance peak.

Resonant technology is currently being widely incorporated in modern inductive wireless power systems. One of the possibilities envisioned for this technology is area wireless power coverage. A coil in the wall or ceiling of a room might be able to wirelessly power lights and mobile devices anywhere in the room, with reasonable efficiency. An environmental and economic benefit of wirelessly powering small devices such as clocks, radios, music players and remote controls is that it could drastically reduce the 6 billion batteries disposed of each year, a large source of toxic waste and groundwater contamination.

Capacitive Coupling

In capacitive coupling (electrostatic induction), the conjugate of inductive coupling, energy is transmitted by electric fields between electrodes such as metal plates. The transmitter and receiver electrodes form a capacitor, with the intervening space as the dielectric. An alternating voltage generated by the transmitter is applied to the transmitting plate, and the oscillating electric field induces an alternating potential on the receiver plate by electrostatic induction, which causes an alternating current to flow in the load circuit. The amount of power transferred increases with the frequency the square of the voltage, and the capacitance between the plates, which is proportional to the area of the smaller plate and (for short distances) inversely proportional to the separation.

Capacitive coupling has only been used practically in a few low power applications, because the very high voltages on the electrodes required to transmit significant power can be hazardous, and can cause unpleasant side effects such as noxious ozone production. In addition, in contrast to magnetic fields, electric fields interact strongly with most materials, including the human body, due to dielectric polarization. Intervening materials between or near the electrodes can absorb the energy, in the case of humans possibly causing excessive electromagnetic field exposure. However capacitive coupling has a few advantages over inductive coupling. The field is largely confined between the capacitor plates, reducing interference, which in inductive coupling requires heavy ferrite "flux confinement" cores. Also, alignment requirements between the transmitter and receiver are less critical. Capacitive coupling has recently been applied to charging battery powered portable devices. and is being considered as a means of transferring power between substrate layers in integrated circuits.

Capacitive Wireless Power Systems

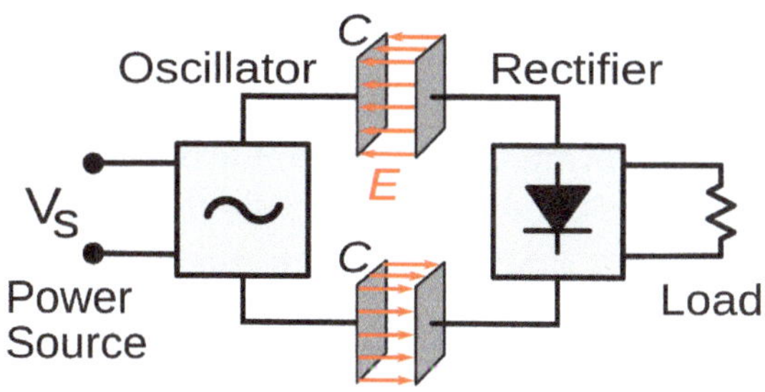

transverse or bipolar coupling

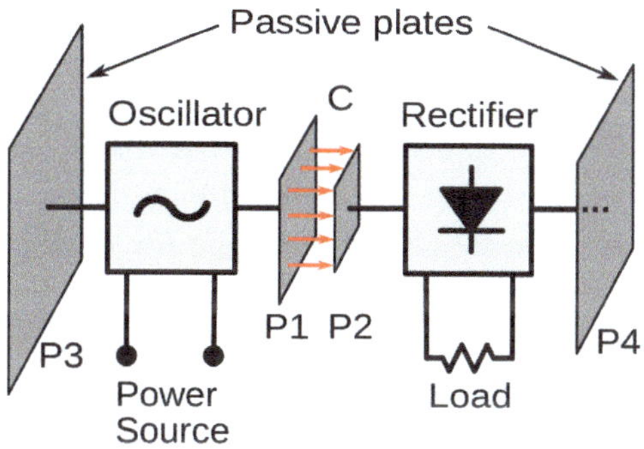

longitudinal or unipolar coupling

Two types of circuit have been used:

- Bipolar design: In this type of circuit, there are two transmitter plates and two receiver plates. Each transmitter plate is coupled to a receiver plate. The transmitter oscillator drives the transmitter plates in opposite phase (180° phase difference) by a high alternating voltage, and the load is connected between the two receiver plates. The alternating electric fields induce opposite phase alternating potentials in the receiver plates, and this "push-pull" action causes current to flow back and forth between the plates through the load. A disadvantage of this configuration for wireless charging is that the two plates in the receiving device must be aligned face to face with the charger plates for the device to work.
- Unipolar design: In this type of circuit, the transmitter and receiver have only one active electrode, and either the ground or a large passive electrode serves as the return path for the current. The transmitter oscillator is connected between an active and a passive electrode. The load is also connected between an active and a passive electrode. The electric field produced by the transmitter induces alternating charge displacement in the load dipole through electrostatic induction.

Resonant Capacitive Coupling

Resonance can also be used with capacitive coupling to extend the range. At the turn of the 20th century, Nikola Tesla did the first experiments with both resonant electrostatic and magnetic coupling.

Electrical Conduction

Atmospheric Plasma Channel Coupling

In atmospheric plasma channel coupling, energy is transferred between two electrodes by electrical conduction through ionized air. When an electric field gradient exists between the two electrodes, exceeding 34 kilovolts per centimeter at sea level atmospheric pressure, an electric arc occurs. This atmospheric dielectric breakdown results in the flow of electric current along a random trajectory through an ionized plasma channel between the two electrodes. An example of this is natural lightning, where one electrode is a virtual point in a cloud and the other is a point on Earth. Laser Induced Plasma Channel (LIPC) research is presently underway using ultrafast lasers to artificially promote development of the plasma channel through the air, directing the electric arc, and guiding the current across a specific path in a controllable manner. The laser energy reduces the atmospheric dielectric breakdown voltage and the air is made less insulating by superheating, which lowers the density (p) of the filament of air.

This new process is being explored for use as a laser lightning rod and as a means to trigger lightning bolts from clouds for natural lightning channel studies, for artificial atmospheric propagation studies, as a substitute for conventional radio antennas, for applications associated with electric welding and machining, for diverting power from high-voltage capacitor discharges, for directed-energy weapon applications employing electrical conduction through a ground return path, and electronic jamming.

Magnetodynamic Coupling

In this method, power is transmitted between two rotating armatures, one in the transmitter and one in the receiver, which rotate synchronously, coupled together by a magnetic field generated by permanent magnets on the armatures. The transmitter armature is turned either by or as the rotor of an electric motor, and its magnetic field exerts torque on the receiver armature, turning it. The magnetic field acts like a mechanical coupling between the armatures. The receiver armature produces power to drive the load, either by turning a separate electric generator or by using the receiver armature itself as the rotor in a generator.

This device has been proposed as an alternative to inductive power transfer for noncontact charging of electric vehicles. A rotating armature embedded in a garage floor or curb would turn a receiver armature in the underside of the vehicle to charge its batteries. It is claimed that this technique can transfer power over distances of 10 to 15 cm (4 to 6 inches) with high efficiency, over 90%. Also, the low frequency stray magnetic fields produced by the rotating magnets produce less electromagnetic interference to nearby electronic devices than the high frequency magnetic fields produced by inductive coupling systems. A prototype system charging electric vehicles has been in operation at University of British Columbia since 2012. Other researchers, however, claim that the two energy conversions (electrical to mechanical to electrical again) make the system less efficient than electrical systems like inductive coupling.

Far-field Radiative Techniques

Far field methods achieve longer ranges, often multiple kilometer ranges, where the distance is much greater than the diameter of the device(s). The main reason for longer ranges with radio wave and optical devices is the fact that electromagnetic radiation in the far-field can be made to match the shape of the receiving area (using high directivity antennas or well-collimated laser beams). The maximum directivity for antennas is physically limited by diffraction.

In general, visible light (from lasers) and microwaves (from purpose-designed antennas) are the forms of electromagnetic radiation best suited to energy transfer.

The dimensions of the components may be dictated by the distance from transmitter to receiver, the wavelength and the Rayleigh criterion or diffraction limit, used in standard radio frequency antenna design, which also applies to lasers. Airy's diffraction limit is also frequently used to determine an approximate spot size at an arbitrary distance from the aperture. Electromagnetic radiation experiences less diffraction at shorter wavelengths (higher frequencies); so, for example, a blue laser is diffracted less than a red one.

The Rayleigh criterion dictates that any radio wave, microwave or laser beam will spread and become weaker and diffuse over distance; the larger the transmitter antenna or laser aperture compared to the wavelength of radiation, the tighter the beam and the less it will spread as a function of distance (and vice versa). Smaller antennae also suffer from excessive losses due to side lobes. However, the concept of laser aperture considerably differs from an antenna. Typically, a laser aperture much larger than the wavelength induces multi-moded radiation and mostly collimators are used before emitted radiation couples into a fiber or into space.

Ultimately, beamwidth is physically determined by diffraction due to the dish size in relation to the wavelength of the electromagnetic radiation used to make the beam.

Microwave power beaming can be more efficient than lasers, and is less prone to atmospheric attenuation caused by dust or water vapor.

Here, the power levels are calculated by combining the above parameters together, and adding in the gains and losses due to the antenna characteristics and the transparency and dispersion of the medium through which the radiation passes. That process is known as calculating a link budget.

Microwaves

Power transmission via radio waves can be made more directional, allowing longer distance power beaming, with shorter wavelengths of electromagnetic radiation, typically in the microwave range. A rectenna may be used to convert the microwave energy back into electricity. Rectenna conversion efficiencies exceeding 95% have been realized. Power beaming using microwaves has been proposed for the transmission of energy from orbiting solar power satellites to Earth and the beaming of power to spacecraft leaving orbit has been considered.

An artist's depiction of a solar satellite that could send electric energy by microwaves to a space vessel or planetary surface.

Power beaming by microwaves has the difficulty that, for most space applications, the required aperture sizes are very large due to diffraction limiting antenna directionality. For example, the 1978 NASA Study of solar power satellites required a 1-km diameter transmitting antenna and a 10 km diameter receiving rectenna for a microwave beam at 2.45 GHz. These sizes can be somewhat decreased by using shorter wavelengths, although short wavelengths may have difficulties with atmospheric absorption and beam blockage by rain or water droplets. Because of the "thinned array curse," it is not possible to make a narrower beam by combining the beams of several smaller satellites.

For earthbound applications, a large-area 10 km diameter receiving array allows large total power levels to be used while operating at the low power density suggested for human electromagnetic exposure safety. A human safe power density of 1 mW/cm^2 distributed across a 10 km diameter area corresponds to 750 megawatts total power level. This is the power level found in many modern electric power plants.

Following World War II, which saw the development of high-power microwave emitters known as cavity magnetrons, the idea of using microwaves to transfer power was researched. By 1964, a miniature helicopter propelled by microwave power had been demonstrated.

Japanese researcher Hidetsugu Yagi also investigated wireless energy transmission using a directional array antenna that he designed. In February 1926, Yagi and his colleague Shintaro Uda published their first paper on the tuned high-gain directional array now known as the Yagi antenna. While it did not prove to be particularly useful for power transmission, this beam antenna has been widely adopted throughout the broadcasting and wireless telecommunications industries due to its excellent performance characteristics.

Wireless high power transmission using microwaves is well proven. Experiments in the tens of kilowatts have been performed at Goldstone in California in 1975 and more recently (1997) at Grand Bassin on Reunion Island. These methods achieve distances on the order of a kilometer.

Under experimental conditions, microwave conversion efficiency was measured to be around 54%.

A change to 24 GHz has been suggested as microwave emitters similar to LEDs have been made with very high quantum efficiencies using negative resistance, i.e., Gunn or IMPATT diodes, and this would be viable for short range links.

In 2015, researchers at the University of Washington introduced power over Wi-Fi, which trickle-charges batteries and powered battery-free cameras and temperature sensors using transmissions from Wi-Fi routers. Wi-Fi signals were shown to power battery-free temperature and camera sensors at ranges of up to 20 feet. It was also shown that Wi-Fi can be used to wirelessly trickle-charge nickel–metal hydride and lithium-ion coin-cell batteries at distances of up to 28 feet.

Lasers

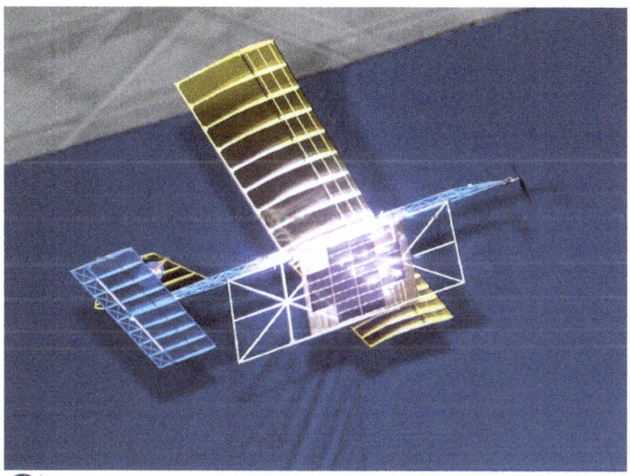

NASA Dryden Flight Research Center Photo Collection
http://www.dfrc.nasa.gov/Gallery/Photo/index.html
NASA Photo: ED03–0249–18 Date: September 18, 2003 Photo By: Tom Tschida

With a laser beam centered on its panel of photovoltaic cells, a model plane makes the first flight of an aircraft powered by a laser beam inside a building at NASA Marshall.

With a laser beam centered on its panel of photovoltaic cells, a lightweight model plane makes the first flight of an aircraft powered by a laser beam inside a building at NASA Marshall Space Flight Center.

In the case of electromagnetic radiation closer to the visible region of the spectrum (tens of micrometers to tens of nanometers), power can be transmitted by converting electricity into a laser beam that is then pointed at a photovoltaic cell. This mechanism is generally known as 'power beaming' because the power is beamed at a receiver that can convert it to electrical energy. At the receiver, special photovoltaic laser power converters which are optimized for monochromatic light conversion are applied.

Advantages compared to other wireless methods are:

- Collimated monochromatic wavefront propagation allows narrow beam cross-section area for transmission over large distances.
- Compact size: solid state lasers fit into small products.
- No radio-frequency interference to existing radio communication such as Wi-Fi and cell phones.
- Access control: only receivers hit by the laser receive power.

Drawbacks include:

- Laser radiation is hazardous. Low power levels can blind humans and other animals. High power levels can kill through localized spot heating.
- Conversion between electricity and light is limited. Photovoltaic cells achieve 40%–50% efficiency. (Note that the conversion efficiency of laser light into electricity is much higher than that of sun light into electricity using solar cells).
- Atmospheric absorption, and absorption and scattering by clouds, fog, rain, etc., causes up to 100% losses.
- Requires a direct line of sight with the target. (Note that, instead of being beamed directly onto the receiver, the laser light can also be guided by an optical fiber. Then one speaks of power-over-fiber technology.)

Laser 'powerbeaming' technology was explored in military weapons and aerospace applications. Also, it is applied for powering of various kinds of sensors in industrial environment. Lately, it is developed for powering commercial and consumer electronics. Wireless energy transfer systems using lasers for consumer space have to satisfy laser safety requirements standardized under IEC 60825.

Other details include propagation, and the coherence and the range limitation problem.

Geoffrey Landis is one of the pioneers of solar power satellites and laser-based transfer of energy especially for space and lunar missions. The demand for safe and frequent space missions has resulted in proposals for a laser-powered space elevator.

NASA's Dryden Flight Research Center demonstrated a lightweight unmanned model plane powered by a laser beam. This proof-of-concept demonstrates the feasibility of periodic recharging using the laser beam system.

Energy Harvesting

In the context of wireless power, *energy harvesting*, also called *power harvesting* or *energy scavenging*, is the conversion of ambient energy from the environment to electric power, mainly to power small autonomous wireless electronic devices. The ambient energy may come from stray electric or magnetic fields or radio waves from nearby electrical equipment, light, thermal energy (heat), or kinetic energy such as vibration or motion of the device. Although the efficiency of conversion is usually low and the power gathered often minuscule (milliwatts or microwatts), it can be adequate to run or recharge small micropower wireless devices such as remote sensors, which are proliferating in many fields. This new technology is being developed to eliminate the need for battery replacement or charging of such wireless devices, allowing them to operate completely autonomously.

History

19th Century Developments

The 19th century saw many developments in mathematics, theories, and counter theories on how electrical energy could be transmitted. In 1826 André-Marie Ampère developed Ampère's circuital law showing that electric current produces a magnetic field. Michael Faraday developed Faraday's law of induction in 1831, describing the electromagnetic force induced in a conductor by a time-varying magnetic flux. The fact that electrical energy could by transmitted at a distance without wires was actually observed by many inventors and experimenters but limited understanding of the phenomenon meant it was usually attributed it to electromagnetic induction. The explanation of these phenomenon would come from the 1860s development of electromagnetism, a theory that unified electricity and magnetism, thought of as two separate forces, with James Clerk Maxwell developing his Maxwell's equations with his prediction of the existence of electromagnetic waves. This was followed on by Heinrich Rudolf Hertz' 1888 proof of the theory, which included the discovery of radio waves.

During the same period two schemes of wireless signaling were put forward by William Henry Ward (1871) and Mahlon Loomis (1872) that were based on the erroneous belief that there was an electrified atmospheric stratum accessible at low altitude. Both inventors patents noted this layer connected with a return path using "Earth currents"' would allow for wireless telegraphy as well as supply power for the telegraph, doing away with artificial batteries, and could also be used for lighting, heat, and motive power. A more piratical demonstrations of wireless transmission via conduction came in Amos Dolbear's 1879 magneto electric telephone that used ground conduction to transmit over a distance of a quarter of a mile.

In 1884 John Henry Poynting developed equations for the flow of power in an electromagnetic field, Poynting's theorem and the Poynting vector, which would be used in the analysis of wireless energy transfer systems.

Tesla

Tesla demonstrating wireless transmission by "electrostatic induction" during an 1891 lecture at Columbia College. The two metal sheets are connected to a Tesla coil oscillator, which applies high-voltage radio frequency alternating current. An oscillating electric field between the sheets ionizes the low-pressure gas in the two long Geissler tubes in his hands, causing them to glow, similar to neon tubes.

After 1890 inventor Nikola Tesla experimented with transmitting power by inductive and capacitive coupling using spark-excited radio frequency resonant transformers, now called Tesla coils, which generated high AC voltages. Early on he attempted to develop a wireless lighting system based on near-field inductive and capacitive coupling and conducted a series of public demonstrations where he lit

Geissler tubes and even incandescent light bulbs from across a stage. He found he could increase the distance at which he could light a lamp by using a receiving LC circuit tuned to resonance with the transmitter's LC circuit. using resonant inductive coupling. Tesla failed to make a commercial product out of his findings but the resonant inductive coupling used is now a familiar technology used throughout electronics and is currently being widely applied to short-range wireless power systems.

(left) Experiment in resonant inductive transfer by Tesla at Colorado Springs 1899. The coil is in resonance with Tesla's magnifying transmitter nearby, powering the light bulb at bottom. *(right)* Tesla's unsuccessful Wardenclyffe power station.

Tesla went on to develop a wireless power distribution system that he hoped would be capable of transmit power long distance directly into homes and factories. Early on he seemed to borrow from the ideas of Mahlon Loomis, proposing a system composed of balloons to suspend transmitting and receiving electrodes in the air above 30,000 feet (9,100 m) in altitude, where he thought the pressure would allow him to send high voltages (millions of volts) long distances. To further study the conductive nature of low pressure air he set up a test facility at high altitude in Colorado Springs during 1899. Experiments he conducted there with a large coil operating in the megavolts range, as well as observations he made of the electronic noise of lightning strikes, led him to incorrectly conclude that he could use the entire globe of the Earth to conduct electrical energy. The theory included driving alternating current pulses into the Earth at its resonant frequency from a grounded Tesla coil working against an elevated capacitance to make the potential of the Earth oscillate. Tesla thought this would allowing alternating current to be received with a similar capacitive antenna tuned to resonance it at any point on Earth with very little power loss. His observations also led him to believe a high voltage used in a coil at an elevation of a few hundred feet would "break the air stratum down", eliminating the need for miles of cable hanging on balloons to create his atmospheric return circuit. Tesla would go on the next year to propose a "World Wireless System" that was to broadcast both information and power worldwide and attempted in 1901 to construct a large high-voltage wireless power station, now called the Wardenclyffe Tower, at Shoreham, New York. By 1904 investment dried up and the facility was never completed.

Near-field and Non-radiative Technologies

Magneto-inductive power transfer between nearby wire coils is old technology, existing since the transformer was developed in the 1800s. Induction heating has been used since the early 1900s.

With the advent of cordless devices, induction charging stands have been developed for appliances used in wet environments, like electric toothbrushes and electric razors, to eliminate the hazard of electric shock. One field to which inductive power transfer has been applied is to power electric vehicles. In 1892 Maurice Hutin and Maurice Leblanc patented a wireless method of powering railroad trains using resonant coils inductively coupled to a track wire at 3 kHz. The first passive RFID (Radio Frequency Identification) technologies were invented by Mario Cardullo (1973) and Koelle et al. (1975) and by the 1990s were being used in proximity cards and contactless smartcards.

The proliferation of portable wireless communication devices such as mobile phones, tablet, and laptop computers in recent decades is currently driving the development of mid-range wireless powering and charging technology to eliminate the need for these devices to be tethered to wall plugs during charging. The Wireless Power Consortium was established in 2008 to develop interoperable standards across manufacturers. Its Qi inductive power standard published in August 2009 enables high efficiency charging and powering of portable devices of up to 5 watts over distances of 4 cm (1.6 inches). The wireless device is placed on a flat charger plate (which can be embedded in table tops at cafes, for example) and power is transferred from a flat coil in the charger to a similar one in the device.

In 2007, a team led by Marin Soljačić at MIT used a dual resonance transmitter with a 25 cm diameter secondary tuned to 10 MHz to transfer 60 W of power to a similar dual resonance receiver over a distance of 2 meters (6.6 ft) (eight times the transmitter coil diameter) at around 40% efficiency. In 2008 the team of Greg Leyh and Mike Kennan of Nevada Lightning Lab used a grounded dual resonance transmitter with a 57 cm diameter secondary tuned to 60 kHz and a similar grounded dual resonance receiver to transfer power through coupled electric fields with an earth return circuit over a distance of 12 meters (39 ft). The Lightning Foundry Twin Coil Prototype coupling distance was more than 21 times the transmitter coil diameter.

Microwaves and Lasers

Before World War 2, little progress was made in wireless power transmission. Radio was developed for communication uses, but couldn't be used for power transmission due to the fact that the relatively low-frequency radio waves spread out in all directions and little energy reached the receiver. In radio communication, at the receiver, an amplifier intensifies a weak signal using energy from another source. For power transmission, efficient transmission required transmitters that could generate higher-frequency microwaves, which can be focused in narrow beams towards a receiver.

The development of microwave technology during World War 2, such as the klystron and magnetron tubes and parabolic antennas made radiative (far-field) methods practical for the first time, and the first long-distance wireless power transmission was achieved in the 1960s by William C. Brown. In 1964 Brown invented the rectenna which could efficiently convert microwaves to DC power, and in 1964 demonstrated it with the first wireless-powered aircraft, a model helicopter powered by microwaves beamed from the ground. A major motivation for microwave research in the 1970s and 80s was to develop a solar power satellite. Conceived in 1968 by Peter Glaser, this would harvest energy from sunlight using solar cells and beam it down to Earth as microwaves to huge rectennas, which would convert it to electrical energy on the electric power grid. In landmark 1975 experiments as technical director of a JPL/Raytheon program, Brown demonstrated long range transmission by beaming 475 W of microwave power to a rectenna a mile away, with a microwave to DC conversion efficiency of 54%. At NASA's Jet Propulsion Laboratory he and Robert Dickinson transmitted 30 kW DC output power across 1.5 km with 2.38 GHz microwaves from a 26 m dish to a 7.3 x 3.5 m rectenna array. The incident-RF to DC conversion efficiency of the rectenna was 80%. In 1983 Japan launched MINIX (Microwave Ionosphere Nonlinear Interaction Experiment), a rocket experiment to test transmission of high power microwaves through the ionosphere.

In recent years a focus of research has been the development of wireless-powered drone aircraft, which began in 1959 with the Dept. of Defense's RAMP (Raytheon Airborne Microwave Platform) project which sponsored Brown's research. In 1987 Canada's Communications Research Center developed a small prototype airplane called Stationary High Altitude Relay Platform (SHARP) to relay telecommunication data between points on earth similar to a communication satellite. Powered by a rectenna, it could fly at 13 miles (21 km) altitude and stay aloft for months. In 1992 a team at Kyoto University built a more advanced craft called MILAX (MIcrowave Lifted Airplane eXperiment).

In 2003 NASA flew the first laser powered aircraft. The small model plane's motor was powered by electricity generated by photocells from a beam of infrared light from a ground-based laser, while a control system kept the laser pointed at the plane.

Wi-Fi

Wi-Fi or WiFi is a technology that allows electronic devices to connect to a wireless LAN (WLAN), mainly using the 2.4 gigahertz (12 cm) UHF and 5 gigahertz (6 cm) SHF ISM radio bands. A WLAN is usually password protected, but may be open, which allows any device within its range to access the resources of the WLAN network.

The Wi-Fi Alliance defines Wi-Fi as any "wireless local area network" (WLAN) product based on the Institute of Electrical and Electronics Engineers' (IEEE) 802.11 standards. "Wi-Fi" is a trademark of the Wi-Fi Alliance. The "Wi-Fi Certified" trademark can only be used by Wi-Fi products that successfully complete Wi-Fi Alliance interoperability certification testing.

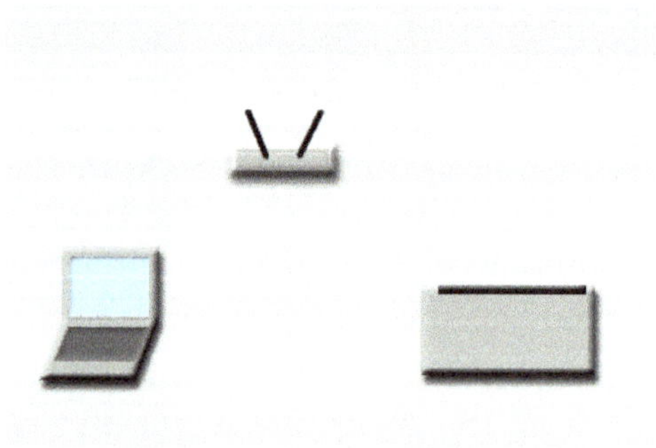

Depiction of a device sending information wirelessly to another device, both connected to the local network, in order to print a document.

Devices that can use Wi-Fi technology include personal computers, video-game consoles, smartphones, digital cameras, tablet computers, digital audio players and modern printers. Wi-Fi compatible devices can connect to the Internet via a WLAN network and a wireless access point. Such an access point (or hotspot) has a range of about 20 meters (66 feet) indoors and a greater range outdoors. Hotspot coverage can be as small as a single room with walls that block radio waves, or as large as many square kilometres achieved by using multiple overlapping access points.

Wi-Fi is less secure than wired connections, such as Ethernet, precisely because an intruder does not need a physical connection. Web pages that use TLS are secure, but unencrypted Internet access can easily be detected by intruders. Because of this, Wi-Fi has adopted various encryption technologies. The early encryption WEP proved easy to break. Higher quality protocols (WPA, WPA2) were added later. An optional feature added in 2007, called Wi-Fi Protected Setup (WPS), had a serious flaw that allowed an attacker to recover the router's password. The Wi-Fi Alliance has since updated its test plan and certification program to ensure all newly certified devices resist attacks.

History

In 1971, ALOHAnet connected the Hawaiian Islands with a UHF wireless packet network. ALOHAnet and the ALOHA protocol were early forerunners to Ethernet, and later the IEEE 802.11 protocols, respectively.

A 1985 ruling by the U.S. Federal Communications Commission released the ISM band for unlicensed use. These frequency bands are the same ones used by equipment such as microwave ovens and are subject to interference.

In 1991, NCR Corporation with AT&T Corporation invented the precursor to 802.11, intended for use in cashier systems. The first wireless products were under the name WaveLAN. They are the ones credited with inventing Wi-Fi.

The Australian radio-astronomer Dr John O'Sullivan with his colleagues Dr Terrence Percival AM, Mr Graham Daniels, Mr Diet Ostry, Mr John Deane developed a key patent used in Wi-Fi as a by-product of a Commonwealth Scientific and Industrial Research Organisation (CSIRO) research project, "a failed experiment to detect exploding mini black holes the size of an atomic particle". In 1992 and 1996, CSIRO obtained patents for a method later used in Wi-Fi to "unsmear" the signal.

The first version of the 802.11 protocol was released in 1997, and provided up to 2 Mbit/s link speeds. This was updated in 1999 with 802.11b to permit 11 Mbit/s link speeds, and this proved to be popular.

In 1999, the Wi-Fi Alliance formed as a trade association to hold the Wi-Fi trademark under which most products are sold.

Wi Fi uses a large number of patents held by many different organizations. In April 2009, 14 technology companies agreed to pay CSIRO $250 million for infringements on CSIRO patents. This led to Australia labeling Wi-Fi as an Australian invention, though this has been the subject of some controversy. CSIRO won a further $220 million settlement for Wi-Fi patent-infringements in 2012 with global firms in the United States required to pay the CSIRO licensing rights estimated to be worth an additional $1 billion in royalties. In 2016, the wireless local area network Test Bed was chosen as Australia's contribution to the exhibition *A History of the World in 100 Objects* held in the National Museum of Australia. In high internet cost countries such as South Africa, government allocating wifi spectrum to wireless internet service providers have helped reduce internet prices as these smaller companies are able to deliver high speed broadband at a much lower cost than the big telcos.

Etymology

The name *Wi-Fi*, commercially used at least as early as August 1999, was coined by the brand-consulting firm Interbrand. The Wi-Fi Alliance had hired Interbrand to create a name that was "a little catchier than 'IEEE 802.11b Direct Sequence.' " Phil Belanger, a founding member of the Wi-Fi Alliance who presided over the selection of the name "Wi-Fi," has stated that Interbrand invented *Wi-Fi* as a pun upon the word *hi-fi*

Interbrand also created the Wi-Fi logo. The yin-yang Wi-Fi logo indicates the certification of a product for interoperability.

The Wi-Fi Alliance used the "nonsense" advertising slogan "The Standard for Wireless Fidelity" for a short time after the brand name was created, leading to the misconception that Wi-Fi was an abbreviation of "Wireless Fidelity." The Wi-Fi Alliance was also called the "Wireless Fidelity Alliance Inc" in some publications and the IEEE's own website has stated "WiFi is a short name for Wireless Fidelity".

Non-Wi-Fi technologies intended for fixed points, such as Motorola Canopy, are usually described as fixed wireless. Alternative wireless technologies include mobile phone standards, such as 2G, 3G, 4G, and LTE.

The name is often written as WiFi or Wifi, but these are not approved by the Wi-Fi Alliance.

Wi-Fi Ad Hoc Mode

Wi-Fi nodes operating in ad hoc mode refers to devices talking directly to each other without the need to first talk to an access point (also known as base station). Wi-Fi ad hoc mode was first invented and realized by Chai K Toh in his 1996 invention of Wi-Fi ad hoc routing, implemented on Lucent WaveLAN 802.11a wireless on IBM Think Pads over a 6 nodes scenario spanning a region of over a mile. The success was recorded in Mobile Computing magazine and later published formally in IEEE Transactions on Wireless Communications.

Wi-Fi Certification

The IEEE does not test equipment for compliance with their standards. The non-profit Wi-Fi Alliance was formed in 1999 to fill this void — to establish and enforce standards for interoperability and backward compatibility, and to promote wireless local-area-network technology. As of 2010, the Wi-Fi Alliance consisted of more than 375 companies from around the world. The Wi-Fi Alliance enforces the use of the Wi-Fi brand to technologies based on the IEEE 802.11 standards from the IEEE. This includes wireless local area network (WLAN) connections, device to device connectivity (such as Wi-Fi Peer to Peer aka Wi-Fi Direct), Personal area network (PAN), local area network (LAN) and even some limited wide area network (WAN) connections. Manufacturers with membership in the Wi-Fi Alliance, whose products pass the certification process, gain the right to mark those products with the Wi-Fi logo.

Specifically, the certification process requires conformance to the IEEE 802.11 radio standards, the WPA and WPA2 security standards, and the EAP authentication standard. Certification may optionally include tests of IEEE 802.11 draft standards, interaction with cellular-phone technology in converged devices, and features relating to security set-up, multimedia, and power-saving.

Not every Wi-Fi device is submitted for certification. The lack of Wi-Fi certification does not necessarily imply that a device is incompatible with other Wi-Fi devices. The Wi-Fi Alliance may or may not sanction derivative terms, such as Super Wi-Fi, coined by the US Federal Communications Commission (FCC) to describe proposed networking in the UHF TV band in the US.

IEEE 802.11 Standard

This Netgear Wi-Fi router contains dual bands for transmitting the 802.11 standard across the 2.4 and 5 GHz spectrums.

The IEEE 802.11 standard is a set of media access control (MAC) and physical layer (PHY) specifications for implementing wireless local area network (WLAN) computer communication in the 2.4, 3.6, 5, and 60 GHz frequency bands. They are created and maintained by the IEEE LAN/MAN Standards Committee (IEEE 802). The base version of the standard was released in 1997, and has had subsequent amendments. The standard and amendments provide the basis for wireless network products using the Wi-Fi brand. While each amendment is officially revoked when it is incorporated in the latest version of the standard, the corporate world tends to market to the revisions because they concisely denote capabilities of their products. As a result, in the market place, each revision tends to become its own standard.

Uses

A Japanese sticker indicating to the public that a location is within range of a Wi-Fi network. A dot with curved lines radiating from it is a common symbol for Wi-Fi, representing a point transmitting a signal.

To connect to a Wi-Fi LAN, a computer has to be equipped with a wireless network interface controller. The combination of computer and interface controller is called a *station*. For all stations that share a single radio frequency communication channel, transmissions on this channel are received by all stations within range. The transmission is not guaranteed to be delivered and is therefore a best-effort delivery mechanism. A carrier wave is used to transmit the data. The data is organised in packets on an Ethernet link, referred to as "Ethernet frames".

Internet Access

Wi-Fi technology may be used to provide Internet access to devices that are within the range of a wireless network that is connected to the Internet. The coverage of one or more interconnected access points (*hotspots*) can extend from an area as small as a few rooms to as large as many square kilometres. Coverage in the larger area may require a group of access points with overlapping coverage. For example, public outdoor Wi-Fi technology has been used successfully in wireless mesh networks in London, UK. An international example is Fon.

Wi-Fi provides service in private homes, businesses, as well as in public spaces at Wi-Fi hotspots set up either free-of-charge or commercially, often using a captive portal webpage for access. Organizations and businesses, such as airports, hotels, and restaurants, often provide free-use hotspots to attract customers. Enthusiasts or authorities who wish to provide services or even to promote business in selected areas sometimes provide free Wi-Fi access.

Routers that incorporate a digital subscriber line modem or a cable modem and a Wi-Fi access point, often set up in homes and other buildings, provide Internet access and internetworking to all devices connected to them, wirelessly or via cable.

Similarly, battery-powered routers may include a cellular Internet radiomodem and Wi-Fi access point. When subscribed to a cellular data carrier, they allow nearby Wi-Fi stations to access the Internet over 2G, 3G, or 4G networks using the tethering technique. Many smartphones have a built-in capability of this sort, including those based on Android, BlackBerry, Bada, iOS (iPhone), Windows Phone and Symbian, though carriers often disable the feature, or charge a separate fee to enable it, especially for customers with unlimited data plans. "Internet packs" provide standalone facilities of this type as well, without use of a smartphone; examples include the MiFi- and WiBro-branded devices. Some laptops that have a cellular modem card can also act as mobile Internet Wi-Fi access points.

Wi-Fi also connects places that normally don't have network access, such as kitchens and garden sheds.

City-wide Wi-Fi

In the early 2000s, many cities around the world announced plans to construct citywide Wi-Fi networks. There are many successful examples; in 2004, Mysore became India's first Wi-Fi-enabled city. A company called WiFiyNet has set up hotspots in Mysore, covering the complete city and a few nearby villages.

In 2005, St. Cloud, Florida and Sunnyvale, California, became the first cities in the United States to offer citywide free Wi-Fi (from MetroFi). Minneapolis has generated $1.2 million in profit annually for its provider.

An outdoor Wi-Fi access point

In May 2010, London, UK, Mayor Boris Johnson pledged to have London-wide Wi-Fi by 2012. Several boroughs including Westminster and Islington already had extensive outdoor Wi-Fi coverage at that point.

Officials in South Korea's capital are moving to provide free Internet access at more than 10,000 locations around the city, including outdoor public spaces, major streets and densely populated residential areas. Seoul will grant leases to KT, LG Telecom and SK Telecom. The companies will invest $44 million in the project, which was to be completed in 2015.

Campus-wide Wi-Fi

Many traditional university campuses in the developed world provide at least partial Wi-Fi coverage. Carnegie Mellon University built the first campus-wide wireless Internet network, called Wireless Andrew, at its Pittsburgh campus in 1993 before Wi-Fi branding originated. By February 1997 the CMU wifi zone was fully operational. Many universities collaborate in providing Wi-Fi access to students and staff through the eduroam international authentication infrastructure.

Direct Computer-to-computer Communications

Wi-Fi also allows communications directly from one computer to another without an access point intermediary. This is called *ad hoc* Wi-Fi transmission. This wireless ad hoc network mode has proven popular with multiplayer handheld game consoles, such as the Nintendo DS, PlayStation Portable, digital cameras, and other consumer electronics devices. Some devices can also share their Internet connection using ad hoc, becoming hotspots or "virtual routers".

Similarly, the Wi-Fi Alliance promotes the specification Wi-Fi Direct for file transfers and media sharing through a new discovery- and security-methodology. Wi-Fi Direct launched in October 2010.

Another mode of direct communication over Wi-Fi is Tunneled Direct Link Setup (TDLS), which enables two devices on the same Wi-Fi network to communicate directly, instead of via the access point.

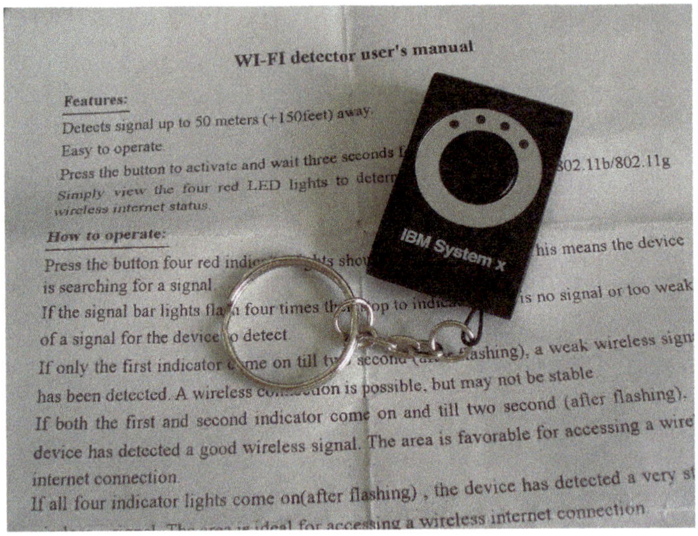

A keychain-size Wi-Fi detector

Wi-Fi Radio Spectrum

802.11b and 802.11g use the 2.4 GHz ISM band, operating in the United States under Part 15 Rules and Regulations. Because of this choice of frequency band, 802.11b and g equipment may occasionally suffer interference from microwave ovens, cordless telephones, and Bluetooth devices.

Spectrum assignments and operational limitations are not consistent worldwide: Australia and Europe allow for an additional two channels (12, 13) beyond the 11 permitted in the United States for the 2.4 GHz band, while Japan has three more (12–14). In the US and other countries, 802.11a and 802.11g devices may be operated without a license, as allowed in Part 15 of the FCC Rules and Regulations.

A Wi-Fi signal occupies five channels in the 2.4 GHz band. Any two channel numbers that differ by five or more, such as 2 and 7, do not overlap. The oft-repeated adage that channels 1, 6, and 11 are the *only* non-overlapping channels is, therefore, not accurate. Channels 1, 6, and 11 are the only *group of three* non-overlapping channels in North America and the United Kingdom. In Europe and Japan using Channels 1, 5, 9, and 13 for 802.11g and 802.11n is recommended.

802.11a uses the 5 GHz U-NII band, which, for much of the world, offers at least 23 non-overlapping channels rather than the 2.4 GHz ISM frequency band, where adjacent channels overlap.

Interference

Wi-Fi connections can be disrupted or the Internet speed lowered by having other devices in the same area. Many 2.4 GHz 802.11b and 802.11g access-points default to the same channel on initial startup, contributing to congestion on certain channels. Wi-Fi pollution, or an excessive number of access points in the area, especially on the neighboring channel, can prevent access and interfere with other devices' use of other access points, caused by overlapping channels in the 802.11g/b spectrum, as well as with decreased signal-to-noise ratio (SNR) between access points. This can become a problem in high-density areas, such as large apartment complexes or office buildings with many Wi-Fi access points. It is advised to only use channel 1-6-11.

Additionally, other devices use the 2.4 GHz band: microwave ovens, ISM band devices, security cameras, ZigBee devices, Bluetooth devices, video senders, cordless phones, baby monitors, and, in some countries, amateur radio, all of which can cause significant additional interference. It is also an issue when municipalities or other large entities (such as universities) seek to provide large area coverage.

Service Set Identifier (SSID)

In addition to running on different channels, multiple Wi-Fi networks can share channels.

A service set is the set of all the devices associated with a particular Wi-Fi network. The service set can be local, independent, extended or mesh.

Each service set has an associated identifier, the 32-byte Service Set Identifier (SSID), which identifies the particular network. The SSID is configured within the devices that are considered part of the network, and it is transmitted in the packets. Receivers ignore wireless packets from networks with a different SSID.

Throughput

As the 802.11 specifications evolved to support higher throughput, the bandwidth requirements also increased to support them. 802.11n uses double the radio spectrum/bandwidth (40 MHz) compared to 802.11a or 802.11g (20 MHz).[76] This means there can be only one 802.11n network on the 2.4 GHz band at a given location, without interference to/from other WLAN traffic. 802.11n can also be set to limit itself to 20 MHz bandwidth to prevent interference in dense community.

Many newer consumer devices support the latest 802.11ac standard, which uses the 5 GHz band exclusively and is capable of multi-station WLAN throughput of at least 1 gigabit per second. According to a study, devices with the 802.11ac specification were expected to be common by 2015 with an estimated one billion spread around the world.

Hardware

Wi-Fi whitelist triggered on an HP laptop

Wi-Fi allows cheaper deployment of local area networks (LANs). Also, spaces where cables cannot be run, such as outdoor areas and historical buildings, can host wireless LANs. However, building walls of certain materials, such as stone with high metal content, can block Wi-Fi signals.

Manufacturers are building wireless network adapters into most laptops. The price of chipsets for Wi-Fi continues to drop, making it an economical networking option included in even more devices.

Different competitive brands of access points and client network-interfaces can inter-operate at a basic level of service. Products designated as "Wi-Fi Certified" by the Wi-Fi Alliance are backwards compatible. Unlike mobile phones, any standard Wi-Fi device will work anywhere in the world.

Standard Devices

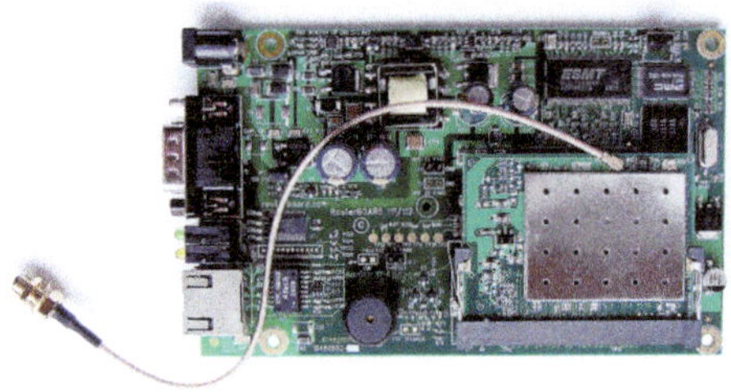

An embedded RouterBoard 112 with U.FL-RSMA pigtail and R52 mini PCI Wi-Fi card widely used by wireless Internet service providers (WISPs) in the Czech Republic

OSBRiDGE 3GN – 802.11n Access Point and UMTS/GSM Gateway in one device

An Atheros draft-N Wi-Fi adapter with built in Bluetooth on a Sony Vaio E series laptop

USB wireless adapter

A wireless access point (WAP) connects a group of wireless devices to an adjacent wired LAN. An access point resembles a network hub, relaying data between connected wireless devices in addition to a (usually) single connected wired device, most often an Ethernet hub or switch, allowing wireless devices to communicate with other wired devices.

Wireless adapters allow devices to connect to a wireless network. These adapters connect to devices using various external or internal interconnects such as PCI, miniPCI, USB, ExpressCard, Cardbus and PC Card. As of 2010, most newer laptop computers come equipped with built in internal adapters.

Wireless routers integrate a Wireless Access Point, Ethernet switch, and internal router firmware application that provides IP routing, NAT, and DNS forwarding through an integrated WAN-interface. A wireless router allows wired and wireless Ethernet LAN devices to connect to a (usually) single WAN device such as a cable modem or a DSL modem. A wireless router allows all three devices, mainly the access point and router, to be configured through one central utility. This utility is usually an integrated web server that is accessible to wired and wireless LAN clients and often optionally to WAN clients. This utility may also be an application that is run on a computer, as is the case with as Apple's AirPort, which is managed with the AirPort Utility on Mac OS X and iOS.

Wireless network bridges connect a wired network to a wireless network. A bridge differs from an access point: an access point connects wireless devices to a wired network at the data-link layer. Two wireless bridges may be used to connect two wired networks over a wireless link, useful in situations where a wired connection may be unavailable, such as between two separate homes or for devices which do not have wireless networking capability (but have wired networking capability), such as consumer entertainment devices; alternatively, a wireless bridge can be used to enable a device which supports a wired connection to operate at a wireless networking standard which is faster than supported by the wireless network connectivity feature (external dongle or inbuilt) supported by the device (e.g. enabling Wireless-N speeds (up to the maximum supported speed on the wired Ethernet port on both the bridge and connected devices including the wireless access point) for a device which only supports Wireless-G). A dual-band wireless bridge can also be used to enable 5 GHz wireless network operation on a device which only supports 2.4 GHz wireless networking functionality and has a wired Ethernet port.

Wireless range-extenders or wireless repeaters can extend the range of an existing wireless network. Strategically placed range-extenders can elongate a signal area or allow for the signal area to reach around barriers such as those pertaining in L-shaped corridors. Wireless devices connected through repeaters will suffer from an increased latency for each hop, as well as from a reduction in the maximum data throughput that is available. In addition, the effect of additional users using a network employing wireless range-extenders is to consume the available bandwidth faster than would be the case where but a single user migrates around a network employing extenders. For this reason, wireless range-extenders work best in networks supporting very low traffic throughput requirements, such as for cases where but a single user with a Wi-Fi equipped tablet migrates around the combined extended and non-extended portions of the total connected network. Additionally, a wireless device connected to any of the repeaters in the chain will have a data throughput that is also limited by the "weakest link" existing in the chain between where the connection originates and where the connection ends. Networks employing wireless extenders are also more prone to degradation from interference from neighboring access points that border portions of the extended network and that happen to occupy the same channel as the extended network.

The security standard, Wi-Fi Protected Setup, allows embedded devices with limited graphical user interface to connect to the Internet with ease. Wi-Fi Protected Setup has 2 configurations: The Push Button configuration and the PIN configuration. These embedded devices are also called The Internet of Things and are low-power, battery-operated embedded systems. A number of Wi-Fi manufacturers design chips and modules for embedded Wi-Fi, such as GainSpan.

Embedded Systems

Increasingly in the last few years (particularly as of 2007), embedded Wi-Fi modules have become available that incorporate a real-time operating system and provide a simple means of wirelessly enabling any device which has and communicates via a serial port. This allows the design of simple monitoring devices. An example is a portable ECG device monitoring a patient at home. This Wi-Fi-enabled device can communicate via the Internet.

Embedded serial-to-Wi-Fi module

These Wi-Fi modules are designed by OEMs so that implementers need only minimal Wi-Fi knowledge to provide Wi-Fi connectivity for their products.

In June 2014 Texas Instruments introduced the first ARM Cortex-M4 microcontroller with an onboard dedicated Wi-Fi MCU, the SimpleLink CC3200. It makes embedded systems with Wi-Fi connectivity possible to build as single-chip devices, which reduces their cost and minimum size, making it more practical to build wireless-networked controllers into inexpensive ordinary objects.

Range

The Wi-Fi signal range depends on the frequency band, radio power output, antenna gain and antenna type as well as the modulation technique. Line-of-sight is the thumbnail guide but reflection and refraction can have a significant impact.

An access point compliant with either 802.11b or 802.11g, using the stock antenna might have a range of 100 m (330 ft). The same radio with an external semi parabolic antenna (15 dB gain) might have a range over 20 miles.

Higher gain rating (dBi) indicates further deviation (generally toward the horizontal) from a theoretical, perfect isotropic radiator, and therefore the further the antenna can project a usable signal, as compared to a similar output power on a more isotropic antenna. For example, an 8 dBi antenna used with a 100 mW driver will have a similar horizontal range to a 6 dBi antenna being driven at 500 mW. Note that this assumes that radiation in the vertical is lost; this may not be the case in some situations, especially in large buildings or within a waveguide. In the above example, a directional waveguide could cause the low power 6 dBi antenna to project much further in a single direction than the 8 dBi antenna which is not in a waveguide, even if they are both being driven at 100 mW.

IEEE 802.11n, however, can more than double the range. Range also varies with frequency band. Wi-Fi in the 2.4 GHz frequency block has slightly better range than Wi-Fi in the 5 GHz frequency block used by 802.11a (and optionally by 802.11n). On wireless routers with detachable antennas, it is possible to improve range by fitting upgraded antennas which have higher gain in particular directions. Outdoor ranges can be improved to many kilometers through the use of high gain directional antennas at the router and remote device(s). In general, the maximum amount of power that a Wi-Fi device can transmit is limited by local regulations, such as FCC Part 15 in the US. Equivalent isotropically radiated power (EIRP) in the European Union is limited to 20 dBm (100 mW).

To reach requirements for wireless LAN applications, Wi-Fi has fairly high power consumption compared to some other standards. Technologies such as Bluetooth (designed to support wireless personal area network (PAN) applications) provide a much shorter propagation range between 1 and 100 m and so in general have a lower power consumption. Other low-power technologies such as ZigBee have fairly long range, but much lower data rate. The high power consumption of Wi-Fi makes battery life in mobile devices a concern.

Researchers have developed a number of "no new wires" technologies to provide alternatives to Wi-Fi for applications in which Wi-Fi's indoor range is not adequate and where installing new wires (such as CAT-6) is not possible or cost-effective. For example, the ITU-T G.hn standard for high speed local area networks uses existing home wiring (coaxial cables, phone lines and power lines). Although G.hn does not provide some of the advantages of Wi-Fi (such as mobility or outdoor use), it is designed for applications (such as IPTV distribution) where indoor range is more important than mobility.

For the best performance, a number of people only recommend using wireless networking as a supplement to wired networking.

Due to the complex nature of radio propagation at typical Wi-Fi frequencies, particularly the effects of signal reflection off trees and buildings, algorithms can only approximately predict Wi-Fi signal strength for any given area in relation to a transmitter. This effect does not apply equally to long-range Wi-Fi, since longer links typically operate from towers that transmit above the surrounding foliage.

The practical range of Wi-Fi essentially confines mobile use to such applications as inventory-taking machines in warehouses or in retail spaces, barcode-reading devices at check-out stands, or receiving/shipping stations. Mobile use of Wi-Fi over wider ranges is limited, for instance, to uses such as in an automobile moving from one hotspot to another. Other wireless technologies are more suitable for communicating with moving vehicles.

Distance Records

Distance records (using non-standard devices) include 382 km (237 mi) in June 2007, held by Ermanno Pietrosemoli and EsLaRed of Venezuela, transferring about 3 MB of data between the mountain-tops of El Águila and Platillon. The Swedish Space Agency transferred data 420 km (260 mi), using 6 watt amplifiers to reach an overhead stratospheric balloon.

Multiple Access Points

Increasing the number of Wi-Fi access points provides network redundancy, better range, support for fast roaming and increased overall network-capacity by using more channels or by defining smaller cells. Except for the smallest implementations (such as home or small office networks), Wi-Fi implementations have moved toward "thin" access points, with more of the network intelligence housed in a centralized network appliance, relegating individual access points to the role of "dumb" transceivers. Outdoor applications may use mesh topologies.

When multiple access points are deployed they are often configured with the same SSID and security settings to form an "extended service set". Wi-Fi client devices will typically connect to the access point that can provide the strongest signal within that service set.

Network Security

The main issue with wireless network security is its simplified access to the network compared to traditional wired networks such as Ethernet. With wired networking, one must either gain access to a building (physically connecting into the internal network), or break through an external firewall. To enable Wi-Fi, one merely needs to be within the range of the Wi-Fi network. Most business networks protect sensitive data and systems by attempting to disallow external access. Enabling wireless connectivity reduces security if the network uses inadequate or no encryption.

An attacker who has gained access to a Wi-Fi network router can initiate a DNS spoofing attack against any other user of the network by forging a response before the queried DNS server has a chance to reply.

Data Security Risks

The most common wireless encryption-standard, Wired Equivalent Privacy (WEP), has been shown to be easily breakable even when correctly configured. Wi-Fi Protected Access (WPA and WPA2) encryption, which became available in devices in 2003, aimed to solve this problem. Wi-Fi access points typically default to an encryption-free (*open*) mode. Novice users benefit from a zero-configuration device that works out-of-the-box, but this default does not enable any wireless security, providing open wireless access to a LAN. To turn security on requires the user to configure the device, usually via a software graphical user interface (GUI). On unencrypted Wi-Fi networks connecting devices can monitor and record data (including personal information). Such networks can only be secured by using other means of protection, such as a VPN or secure Hypertext Transfer Protocol over Transport Layer Security (HTTPS).

Wi-Fi Protected Access encryption (WPA2) is considered secure, provided a strong passphrase is used. A proposed modification to WPA2 is WPA-OTP or WPA3, which stores an on-chip optically generated onetime pad on all connected devices which is periodically updated via strong encryption then hashed with the data to be sent or received. This would be unbreakable using any (even quantum) computer system as the hashed data is essentially random and no pattern can be detected if it is implemented properly. Main disadvantage is that it would need multi-GB storage chips so would be expensive for the consumers.

Securing Methods

A common measure to deter unauthorized users involves hiding the access point's name by disabling the SSID broadcast. While effective against the casual user, it is ineffective as a security method because the SSID is broadcast in the clear in response to a client SSID query. Another method is to only allow computers with known MAC addresses to join the network, but determined eavesdroppers may be able to join the network by spoofing an authorized address.

Wired Equivalent Privacy (WEP) encryption was designed to protect against casual snooping but it is no longer considered secure. Tools such as AirSnort or Aircrack-ng can quickly recover WEP encryption keys. Because of WEP's weakness the Wi-Fi Alliance approved Wi-Fi Protected Access (WPA) which uses TKIP. WPA was specifically designed to work with older equipment usually through a firmware upgrade. Though more secure than WEP, WPA has known vulnerabilities.

The more secure WPA2 using Advanced Encryption Standard was introduced in 2004 and is supported by most new Wi-Fi devices. WPA2 is fully compatible with WPA.

A flaw in a feature added to Wi-Fi in 2007, called Wi-Fi Protected Setup, allows WPA and WPA2 security to be bypassed and effectively broken in many situations. The only remedy as of late 2011 is to turn off Wi-Fi Protected Setup, which is not always possible.

Virtual Private Networks are often used to secure Wi-Fi.

Piggybacking

Piggybacking refers to access to a wireless Internet connection by bringing one's own computer within the range of another's wireless connection, and using that service without the subscriber's explicit permission or knowledge.

During the early popular adoption of 802.11, providing open access points for anyone within range to use was encouraged to cultivate wireless community networks, particularly since people on average use only a fraction of their downstream bandwidth at any given time.

Recreational logging and mapping of other people's access points has become known as wardriving. Indeed, many access points are intentionally installed without security turned on so that they can be used as a free service. Providing access to one's Internet connection in this fashion may breach the Terms of Service or contract with the ISP. These activities do not result in sanctions in most jurisdictions; however, legislation and case law differ considerably across the world. A proposal to leave graffiti describing available services was called warchalking. A Florida court case determined that owner laziness was not to be a valid excuse.

Piggybacking often occurs unintentionally – a technically unfamiliar user might not change the default "unsecured" settings to their access point and operating systems can be configured to connect automatically to any available wireless network. A user who happens to start up a laptop in the vicinity of an access point may find the computer has joined the network without any visible indication. Moreover, a user intending to join one network may instead end up on another one if the latter has a stronger signal. In combination with automatic discovery of other network resources this could possibly lead wireless users to send sensitive data to the wrong middle-man when seeking a destination. For example, a user could inadvertently use an unsecure network to log in to a website, thereby making the login credentials available to anyone listening, if the website uses an unsecure protocol such as plain HTTP without TLS (HTTPS).

An unauthorized user can obtain security information (factory preset passphrase and/or Wi-Fi Protected Setup PIN) from a label on a wireless access point can use this information (or connect by the Wi-Fi Protected Setup pushbutton method) to commit unauthorized and/or unlawful activities.

Safety

The World Health Organization (WHO) says "no health effects are expected from exposure to RF fields from base stations and wireless networks", but notes that they promote research into effects from other RF sources. Although the WHO's International Agency for Research on Cancer (IARC) later classified radiofrequency electromagnetic fields as "possibly carcinogenic to humans (Group 2B)" (a category used when "a causal association is considered credible, but when chance, bias or confounding cannot be ruled out with reasonable confidence"), this was based on risks associated with wireless phone use rather than Wi-Fi networks.

The United Kingdom's Health Protection Agency reported in 2007 that exposure to Wi-Fi for a year results in the "same amount of radiation from a 20-minute mobile phone call".

A review of studies involving 725 people who claimed electromagnetic hypersensitivity, "...suggests that 'electromagnetic hypersensitivity' is unrelated to the presence of EMF, although more research into this phenomenon is required."

Hotspot (Wi-Fi)

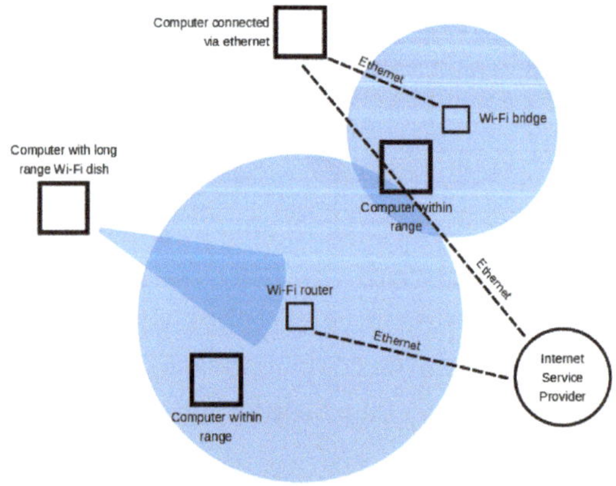

A diagram showing a Wi-Fi network

A hotspot is a physical location where people may obtain Internet access, typically using Wi-Fi technology, via a wireless local area network (WLAN) using a router connected to an internet service provider.

Public hotspots may be found in a number of businesses for use of customers in many developed urban areas throughout the world, such as coffee shops. Many hotels offer wifi access to guests, either in guest rooms or in the lobby. Hotspots differ from wireless access points, which are the hardware devices used to provide a wireless network service. Private hotspots allow Internet access to a device (such as a tablet) via another device which may have data access via say a mobile device.

History

Public park in Brooklyn, New York, has free Wi-Fi from a local corporation

Public access wireless local area networks (LANs) were first proposed by Henrik Sjödin at the NetWorld+Interop conference in The Moscone Center in San Francisco in August 1993. Sjödin did not use the term hotspot but referred to publicly accessible wireless LANs.

The first commercial venture to attempt to create a public local area access network was a firm founded in Richardson, Texas known as PLANCOM (Public Local Area Network Communications). The founders of the venture, Mark Goode, Greg Jackson, and Brett Stewart dissolved the firm in 1998, while Goode and Jackson created MobileStar Networks. The firm was one of the first to sign such public access locations as Starbucks, American Airlines, and Hilton Hotels. The company was sold to Deutsche Telecom in 2001, who then converted the name of the firm into "T-Mobile Hotspot." It was then that the term "hotspot" entered the popular vernacular as a reference to a location where a publicly accessible wireless LAN is available.

ABI Research reported there was a total of 4.9 million global Wi-Fi hotspots in 2012 and projected that number would surpass 6.3 million by the end of 2013. The latest Wireless Broadband Alliance (WBA) Industry Report outlines a positive scenario for the Wi-Fi market: a steady annual increase from 5.2m public hotspots in 2012 to 10.5m public hotspots in 2018. Collectively, WBA operator members serve more than 1 billion subscribers and operate more than 15 million hotspots globally.

Uses

The public can use a laptop or other suitable portable device to access the wireless connection (usually Wi-Fi) provided. Of the estimated 150 million laptops, 14 million PDAs, and other emerging Wi-Fi devices sold per year for the last few years, most include the Wi-Fi feature.

For venues that have broadband Internet access, offering wireless access is as simple as configuring one access point (AP), in conjunction with a router and connecting the AP to the Internet connection. A single wireless router combining these functions may suffice.

The iPass 2014 interactive map, that shows data provided by the analysts Maravedis Rethink, shows that in December 2014 there are 46,000,000 hotspots worldwide and more than 22,000,000 roamable hotspots. More than 10,900 hotspots are on trains, planes and airports (Wi-Fi in motion) and more than 8,500,000 are "branded" hotspots (retail, cafés, hotels). The region with the largest number of public hotspots is Europe, followed by North America and Asia.

Security

Security is a serious concern in connection with Hotspots. There are three possible attack vectors. First, there is the wireless connection between the client and the access point. This needs to be encrypted, so that the connection cannot be eavesdropped or attacked by a man-in-the-middle-attack. Second, there is the Hotspot itself. The WLAN encryption ends at the interface, then travels its network stack unencrypted and then travels over the wired connection up to the BRAS of the ISP. Third, there is the connection from the Access Point to the BRAS of the ISP.

The safest method when accessing the Internet over a Hotspot, with unknown security measures, is end-to-end encryption. Examples of strong end-to-end encryption are HTTPS and SSH.

Locations

Hotspots are often found at airports, bookstores, coffee shops, department stores, fuel stations, hotels, hospitals, libraries, public pay phones, restaurants, RV parks and campgrounds, supermarkets, train stations, and other public places. Additionally, many schools and universities have wireless networks in their campuses.

Types

Free hotspots operate in two ways:

- Using an open public network is the easiest way to create a free hotspot. All that is needed is a Wi-Fi router. Similarly, when users of private wireless routers turn off their authentication requirements, opening their connection, intentionally or not, they permit piggybacking (sharing) by anyone in range.
- Closed public networks use a HotSpot Management System to control access to hotspots. This software runs on the router itself or an external computer allowing operators to authorize only specific users to access the Internet. Providers of such hotspots often associate the free access with a menu, membership, or purchase limit. Operators may also limit each

user's available bandwidth (upload and download speed) to ensure that everyone gets a good quality service. Often this is done through service-level agreements.

Commercial Hotspots

A commercial hotspot may feature:

- A captive portal / login screen / splash page that users are redirected to for authentication and/or payment. The captive portal / splash page sometimes includes the social login buttons.
- A payment option using a credit card, iPass, PayPal, or another payment service (voucher-based Wi-Fi)
- A walled garden feature that allows free access to certain sites
- Service-oriented provisioning to allow for improved revenue
- Data analytics and data capture tools, to analyze and export data from Wi-Fi clients

Many services provide payment services to hotspot providers, for a monthly fee or commission from the end-user income. For example, Amazingports can be used to set up hotspots that intend to offer both fee-based and free internet access, and ZoneCD is a Linux distribution that provides payment services for hotspot providers who wish to deploy their own service.

Major airports and business hotels are more likely to charge for service, though most hotels provide free service to guests; and increasingly, small airports and airline lounges offer free service.. Retail shops, public venues and offices usually provide a free Wi-Fi SSID for their guests and visitors.

Roaming services are expanding among major hotspot service providers. With roaming service the users of a commercial provider can have access to other providers' hotspots, either free of charge or for extra fees, which users will usually be charged on an access-per-minute basis.

Software Hotspots

Many Wi-Fi adapters built into or easily added to consumer computers and mobile devices include the functionality to operate as private or mobile hotspots, sometimes referred to as "mi-fi". The use of a private hotspot to enable other personal devices to access the WAN (usually but not always the Internet) is a form of bridging, and known as tethering. Manufacturers and firmware creators can enable this functionality in Wi-Fi devices on many Wi-Fi devices, depending upon the capabilities of the hardware, and most modern consumer operating systems, including Android, Apple OS X 10.6 and later, Windows mobile, and Linux include features to support this. Additionally wireless chipset manufacturers such as Atheros, Broadcom, Intel and others, may add the capability for certain Wi-Fi NICs, usually used in a client role, to also be used for hotspot purposes. However, some service providers, such as AT&T, Sprint, and T-Mobile charge users for this service or prohibit and disconnect user connections if tethering is detected.

Third-party software vendors offer applications to allow users to operate their own hotspot, whether to access the Internet when on the go, share an existing connection, or extend the range of another hotspot. Third party implementations of software hotspots include:

- AmazingPorts Hotspot software
- Antamedia HotSpot software
- Connectify Hotspot
- Jaze Hotspot Gateway by Jaze Networks
- Hot Spot Network Manager (HSNM)
- Virtual Router
- Tanaza
- Start Hotspot software

Hotspot 2.0

Hotspot 2.0, also known as HS2 and Wi-Fi Certified Passpoint, is an approach to public access Wi-Fi by the Wi-Fi Alliance. The idea is for mobile devices to automatically join a Wi-Fi subscriber service whenever the user enters a Hotspot 2.0 area, in order to provide better bandwidth and services-on-demand to end-users and relieve carrier infrastructure of some traffic.

Hotspot 2.0 is based on the IEEE 802.11u standard, which is a set of protocols published in 2011 to enable cellular-like roaming. If the device supports 802.11u and is subscribed to a Hotspot 2.0 service it will automatically connect and roam.

Supported Devices

- Some Chinese tablet computers
- Some THL smartphones
- Apple mobile devices running iOS 7 and up
- Some Samsung Galaxy smartphones
- Windows 10 devices have full support for network discovery and connection
- Windows 8 and Windows 8.1 lack network discovery, but supports connecting to a network when the credentials are known

Billing

The so-called "User-Fairness-Model " is a dynamic billing model, which allows a volume-based billing, charged only by the amount of payload (data, video, audio). Moreover, the tariff is classified by net traffic and user needs (Pommer, p. 116ff).

If the net traffic increases, then the user has to pay the next higher tariff class. By the way the user is asked for if he still wishes the session also by a higher traffic class. Moreover, in time-critical applications (video, audio) a higher class fare is charged, than for non time-critical applications (such as reading Web pages, e-mail).

<table>
<tr><td colspan="4">Tariff classes of the User-Fairness-Model</td></tr>
<tr><td colspan="2" rowspan="2">low</td><td colspan="2">Net traffic</td></tr>
<tr><td>high</td><td></td></tr>
<tr><td rowspan="2">User needs</td><td>time-critical</td><td>standard</td><td>exclusive</td></tr>
<tr><td>not time-critical</td><td>low priced</td><td>standard</td></tr>
</table>

The "User-fairness model" can be implemented with the help of EDCF (IEEE 802.11e). A EDCF user priority list shares the traffic in 3 access categories (data, video, audio) and user priorities (UP) (Pommer, p. 117):

- Data [UP 0|2]
- Video [UP 5|4]
- Audio [UP 7|6]

Security Concerns

Some hotspots authenticate users; however, this does not prevent users from viewing network traffic using packet sniffers.

Some vendors provide a download option that deploys WPA support. This conflicts with enterprise configurations that have solutions specific to their internal WLAN.

In order to provide robust security to hotspot users, the Wi-Fi Alliance is developing a new hotspot program that aims to encrypt hotspot traffic with WPA2 security. The program was scheduled to launch in the first half of 2012.

Legal Concerns

Depending upon the location, providers of public hotspot access may have legal obligations, related to privacy requirements and liability for use for unlawful purposes. In countries where the internet is regulated or freedom of speech more restricted, there may be requirements such as licensing, logging, or recording of user information. Concerns may also relate to child safety, and social issues such as exposure to objectionable content, protection against cyber-bullying and illegal behaviours, and prevention of perpetration of such behaviors by hotspot users themselves.

European Union

- Data Retention Directive Hotspot owners must retain key user statistics for 12 months.
- Directive on Privacy and Electronic Communications

United Kingdom

- Data Protection Act 1998 The hotspot owner must retain individual's information within the confines of the law.
- Digital Economy Act 2010 Deals with, among other things, copyright infringement, and imposes fines of up to £250,000 for contravention.

Li-Fi

Light Fidelity (Li-Fi) is a bidirectional, high-speed and fully networked wireless communication technology similar to Wi-Fi. The term was coined by Harald Haas and is a form of visible light communication and a subset of optical wireless communications (OWC) and could be a complement to RF communication (Wi-Fi or cellular networks), or even a replacement in contexts of data broadcasting.

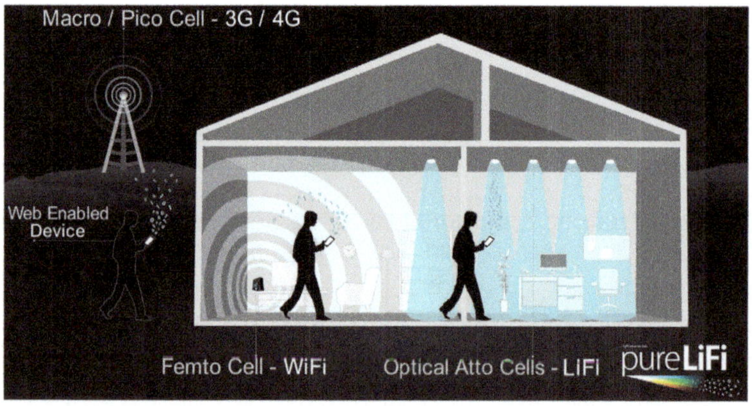

LiFi works in complement with existing and emerging wireless systems.

It is wire and uv visible-light communication or infrared and near-ultraviolet instead of radio-frequency spectrum, part of optical wireless communications technology, which carries much more information, and has been proposed as a solution to the RF-bandwidth limitations.

Technology Details

This OWC technology uses light from light-emitting diodes (LEDs) as a medium to deliver networked, mobile, high-speed communication in a similar manner to Wi-Fi. The Li-Fi market is projected to have a compound annual growth rate of 82% from 2013 to 2018 and to be worth over $6 billion per year by 2018.

Visible light communications (VLC) works by switching the current to the LEDs off and on at a very high rate, too quick to be noticed by the human eye. Although Li-Fi LEDs would have to be kept on to transmit data, they could be dimmed to below human visibility while still emitting enough light to carry data. The light waves cannot penetrate walls which makes a much shorter range, though more secure from hacking, relative to Wi-Fi. Direct line of sight is not necessary for Li-Fi to transmit a signal; light reflected off the walls can achieve 70 Mbit/s.

Li-Fi has the advantage of being useful in electromagnetic sensitive areas such as in aircraft cabins, hospitals and nuclear power plants without causing electromagnetic interference. Both Wi-Fi and Li-Fi transmit data over the electromagnetic spectrum, but whereas Wi-Fi utilizes radio waves, Li-Fi uses visible light. While the US Federal Communications Commission has warned of a potential spectrum crisis because Wi-Fi is close to full capacity, Li-Fi has almost no limitations on capacity. The visible light spectrum is 10,000 times larger than the entire radio frequency spectrum. Researchers have reached data rates of over 10 Gbit/s, which is much faster than typical fast broadband in 2013. Li-Fi is expected to be ten times cheaper than Wi-Fi. Short range, low reliability and high installation costs are the potential downsides.

PureLiFi demonstrated the first commercially available Li-Fi system, the Li-1st, at the 2014 Mobile World Congress in Barcelona.

Bg-Fi is a Li-Fi system consisting of an application for a mobile device, and a simple consumer product, like an IoT (Internet of Things) device, with color sensor, microcontroller, and embedded software. Light from the mobile device display communicates to the color sensor on the consumer product, which converts the light into digital information. Light emitting diodes enable the consumer product to communicate synchronously with the mobile device.

History

Harald Haas, coined the term “Li-Fi” at his TED Global Talk where he introduced the idea of “Wireless data from every light”. He is Chairman of Mobile Communications at the University of Edinburgh and co-founder of pureLiFi.

The general term visible light communication (VLC), whose history dates back to the 1880s, includes any use of the visible light portion of the electromagnetic spectrum to transmit information. The D-Light project at Edinburgh’s Institute for Digital Communications was funded from January 2010 to January 2012. Haas promoted this technology in his 2011 TED Global talk and helped start a company to market it. PureLiFi, formerly pureVLC, is an original equipment manufacturer (OEM) firm set up to commercialize Li-Fi products for integration with existing LED-lighting systems.

In October 2011, companies and industry groups formed the Li-Fi Consortium, to promote high speed optical wireless systems and to overcome the limited amount of radio-based wireless spectrum available by exploiting a completely different part of the electromagnetic spectrum.

A number of companies offer uni-directional VLC products, which is not the same as Li-Fi - a term defined by the IEEE 802.15.7r1 standardization committee.

VLC technology was exhibited in 2012 using Li-Fi. By August 2013, data rates of over 1.6 Gbit/s were demonstrated over a single color LED. In September 2013, a press release said that Li-Fi, or VLC systems in general, do not require line-of-sight conditions. In October 2013, it was reported Chinese manufacturers were working on Li-Fi development kits.

In April 2014, the Russian company Stins Coman announced the development of a Li-Fi wireless local network called BeamCaster. Their current module transfers data at 1.25 gigabytes per second but they foresee boosting speeds up to 5 GB/second in the near future. In 2014 a new record was established by Sisoft (a Mexican company) that was able to transfer data at speeds of up to 10 Gbit/s across a light spectrum emitted by LED lamps.

Standards

Like Wi-Fi, Li-Fi is wireless and uses similar 802.11 protocols; but it uses visible light communication (instead of radio frequency waves), which has much wider bandwidth.

One part of VLC is modeled after communication protocols established by the IEEE 802 workgroup. However, the IEEE 802.15.7 standard is 'out-of-date, it fails to consider the latest technological developments in the field of optical wireless communications, specifically with the introduction of optical orthogonal frequency-division multiplexing (O-OFDM) modulation methods which have been optimized for data rates, multiple-access and energy efficiency. The introduction of O-OFDM means that a new drive for standardization of optical wireless communications is required.

Nonetheless, the IEEE 802.15.7 standard defines the physical layer (PHY) and media access control (MAC) layer. The standard is able to deliver enough data rates to transmit audio, video and multimedia services. It takes into account optical transmission mobility, its compatibility with artificial lighting present in infrastructures, and the interference which may be generated by ambient lighting. The MAC layer permits using the link with the other layers as with the TCP/IP protocol.

The standard defines three PHY layers with different rates:

- The PHY I was established for outdoor application and works from 11.67 kbit/s to 267.6 kbit/s.
- The PHY II layer permits reaching data rates from 1.25 Mbit/s to 96 Mbit/s.
- The PHY III is used for many emissions sources with a particular modulation method called color shift keying (CSK). PHY III can deliver rates from 12 Mbit/s to 96 Mbit/s.

The modulation formats recognized for PHY I and PHY II are on-off keying (OOK) and variable pulse position modulation (VPPM). The Manchester coding used for the PHY I and PHY II layers includes the clock inside the transmitted data by representing a logic 0 with an OOK symbol "01" and a logic 1 with an OOK symbol "10", all with a DC component. The DC component avoids light extinction in case of an extended run of logic 0's.

The first VLC smartphone prototype was presented at the Consumer Electronics Show in Las Vegas from January 7–10 in 2014. The phone uses SunPartner's Wysips CONNECT, a technique that converts light waves into usable energy, making the phone capable of receiving and decoding signals without drawing on its battery. A clear thin layer of crystal glass can be added to small screens like watches and smartphones that make them solar powered. Smartphones could gain 15% more battery life during a typical day. This first smartphones using this technology should arrive in 2015. This screen can also receive VLC signals as well as the smartphone camera. The cost of these screens per smartphone is between $2 and $3, much cheaper than most new technology.

Philips lighting company has developed a VLC system for shoppers at stores. They have to download an app on their smartphone and then their smartphone works with the LEDs in the store. The

LEDs can pinpoint where they are located in the store and give them corresponding coupons and information based on which aisle they are on and what they are looking at.

References

- Dutton, Benjamin (2004). "15 – Basic Radio Navigation". Dutton's Nautical Navigation (15 ed.). Naval Institute Press. pp. 154–163. ISBN 155750248X.
- Watson, Raymond C., Jr. (2009-11-25). Radar Origins Worldwide: History of Its Evolution in 13 Nations Through World War II. Trafford Publishing. ISBN 978-1-4269-2111-7.
- Bush, Stephen F. (2014). Smart Grid: Communication-Enabled Intelligence for the Electric Power Grid. John Wiley & Sons. p. 118. ISBN 1118820231.
- Sun, Tianjia; Xie, Xiang; Zhihua, Wang (2013). Wireless Power Transfer for Medical Microsystems. Springer Science & Business Media. pp. 5–6. ISBN 1461477026.
- Sazonov, Edward; Neuman, Michael R. (2014). Wearable Sensors: Fundamentals, Implementation and Applications. Elsevier. pp. 253–255. ISBN 0124186661.
- Rajakaruna, Sumedha; Shahnia, Farhad; Ghosh, Arindam (2014). Plug In Electric Vehicles in Smart Grids: Integration Techniques. Springer. pp. 34–36. ISBN 981287299X.
- Smith, Glenn S. (1997). An Introduction to Classical Electromagnetic Radiation. Cambridge University Press. p. 474. ISBN 0521586984.
- Tan, Yen Kheng (2013). Energy Harvesting Autonomous Sensor Systems: Design, Analysis, and Practical Implementation. CRC Press. pp. 181–182. ISBN 1439892733.
- Feynman, Richard Phillips; Leighton, Robert B.; Sands, Matthew (1963). The Feynman Lectures on Physics Vol. 1: Mainly Mechanics, Radiation, and Heat. California Institute of Technology. pp. 30.6–30.7. ISBN 0465024939.
- Bolic, Miodrag; Simplot-Ryl, David; Stojmenovic, Ivan (2010). RFID Systems: Research Trends and Challenges. John Wiley & Sons. p. 29. ISBN 0470975660.
- Puers, R. (2008). Omnidirectional Inductive Powering for Biomedical Implants. Springer Science & Business Media. pp. 4–5. ISBN 1402090757.
- Sun, Tianjia; Xie, Xiang; Zhihua, Wang (2013). Wireless Power Transfer for Medical Microsystems. Springer Science & Business Media. ISBN 1461477026.

6

Optical Wireless Communication: An Integrated Study

Light is used to carry information from a distance and this form of communication is known as optical communication. One of the forms of optical communication is optical wireless communications. The aspects elucidated in this section are of vital importance, and provide a better understanding of optical wireless communications.

Optical Wireless Communications

Optical wireless communications (OWC) is a form of optical communication in which unguided visible, infrared (IR), or ultraviolet (UV) light is used to carry a signal.

OWC systems operating in the visible band (390–750 nm) are commonly referred to as visible light communication (VLC). VLC systems take advantage of light emitting diodes (LEDs) which can be pulsed at very high speeds without noticeable effect on the lighting output and human eye. VLC can be possibly used in a wide range of applications including wireless local area networks, wireless personal area networks and vehicular networks among others. On the other hand, terrestrial point-to-point OWC systems, also known as the free space optical (FSO) systems, operate at the near IR frequencies (750–1600 nm). These systems typically use laser transmitters and offer a cost-effective protocol-transparent link with high data rates, i.e., 10 Gbit/s per wavelength, and provide a potential solution for the backhaul bottleneck. There has also been a growing interest on ultraviolet communication (UVC) as a result of recent progress in solid state optical sources/detectors operating within solar-blind UV spectrum (200–280 nm). In this so-called deep UV band, solar radiation is negligible at the ground level and this makes possible the design of photon-counting detectors with wide field-of-view receivers that increase the received energy with little additional background noise. Such designs are particularly useful for outdoor non-line-of-sight configurations to support low power short-range UVC such as in wireless sensor and ad-hoc networks.

History

The proliferation of wireless communications stands out as one of the most significant phenomena in the history of technology. Wireless technologies have become essential much more quickly during the last four decades and they will be a key element of society progress for the foreseeable future. The radio-frequency (RF) technologies wide-scale deployment has become the key factor to the wireless devices and systems expansion. However, the electromagnetic spectrum where the wireless systems are deployed is limited in capacity and costly according to its exclusive licenses of exploitation. With the raise of data heavy wireless communications, the demand for RF spectrum is outstripping supply and they become to consider other viable options for wireless communication using the upper parts of the electromagnetic spectrum not just RF.

Optical wireless communication (OWC) refers to transmission in unguided propagation media through the use of optical carriers, i.e., visible, infrared (IR), and ultraviolet (UV) band. Signalling through beacon fires, smoke, ship flags and semaphore telegraph can be considered the historical forms of OWC. Sunlight has been also used for long distance signalling since very early times. The earliest use of sunlight for communication purposes is attributed to ancient Greeks and Romans who used their polished shields to send signals by reflecting sunlight during battles. In 1810, Carl Friedrich Gauss invented the heliograph which involves a pair of mirrors to direct a controlled beam of sunlight to a distant station. Although the original heliograph was designed for geodetic survey, it was used extensively for military purposes during the late 19th and early 20th century. In 1880, Alexander Graham Bell invented the photophone, known as the world's first wireless telephone system.

The military interest on photophone however continued. For example, in 1935, the German Army developed a photophone where a tungsten filament lamp with an IR transmitting filter was used as a light source. Also, American and German military laboratories continued the development of high pressure arc lamps for optical communication until the 1950s. In modern sense, OWC uses either lasers or light emitting diodes (LEDs) as transmitters. In 1962, MIT Lincoln Labs built an experimental OWC link using a light emitting GaAs diode and was able to transmit TV signals over a distance of 30 miles. After the invention of laser, OWC was envisioned to be the main deployment area for lasers and many trials were conducted using different types of lasers and modulation schemes. However, the results were in general disappointing due to large divergence of laser beams and the inability to cope with atmospheric effects. With the development of low-loss fiber optics in the 1970s, they became the obvious choice for long distance optical transmission and shifted the focus away from OWC systems.

Current Status

Over the decades, the interest in OWC remained mainly limited to covert military applications, and space applications including inter-satellite and deep-space links. OWC's mass market penetration has been so far limited with the exception of IrDA which became a highly successful wireless short-range transmission solution. Development of novel and efficient wireless technologies for a range of transmission links is essential for building future heterogeneous communication networks to support a wide range of service types with various traffic patterns and to meet the ever-increasing demands for higher data rates. Variations of OWC can be potentially employed in a diverse range of communication applications ranging from optical interconnects within integrated circuits through outdoor inter-building links to satellite communications.

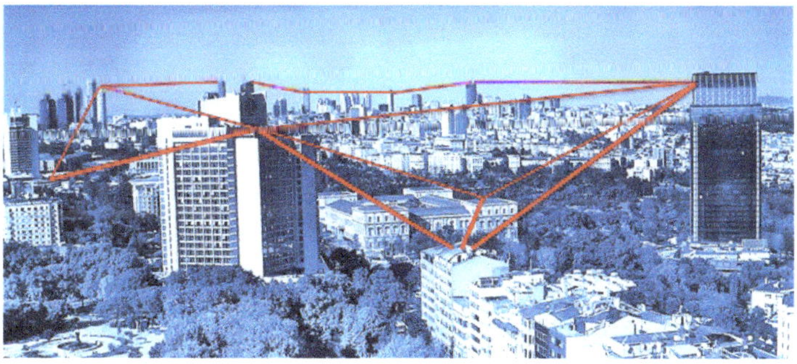

OWC long range inter-buildings communications idea for Istanbul Skyline

Applications

Based on the transmission range, OWC can be studied in five categories:

1. Ultra-short range OWC: chip-to-chip communications in stacked and closely packed multi-chip packages.
2. Short range OWC: wireless body area network (WBAN) and wireless personal area network (WPAN) applications under standard IEEE 802.15.7, underwater communications.
3. Medium range OWC: indoor IR and visible light communications (VLC) for wireless local area networks (WLANs) and inter-vehicular and vehicle-to-infrastructure communications.
4. Long range OWC,: inter-building connections, also called Free-Space Optical Communications (FSO).
5. Ultra-long range OWC: inter-satellite links.

Recent Trends

- In January 2015, IEEE 802.15 formed a Task Group to write a revision to IEEE 802.15.7-2011 that accommodates infrared and near ultraviolet wavelengths, in addition to visible light, and adds options such as Optical Camera Communications and LiFi.
- At long range OWC applications a 1 Gbit/s - 60 km range link between ground to aircraft at 800 km/h speed has been demonstrated, "Extreme Test for the ViaLight Laser Communication Terminal MLT-20 – Optical Downlink from a Jet Aircraft at 800 km/h", DLR and EADS December 2013.
- On consumer devices and short-range OWC applications on phones; Charge and receive data with light at your smartphone: TCL Communication/ALCATEL ONETOUCH and Sunpartner Technologies announces the first fully integrated solar smartphone. March 2014.
- On ultra-long range OWC applications the NASA's Lunar Laser Communication Demonstration (LLCD) transmitted data from lunar orbit to Earth at a rate of 622 Megabits-per-second (Mbps), November 2013.
- The Next Generation of OWC / Visible Light Communications demonstrated 10 Mb/s transmission with Polymer Light-Emitting Diodes or OLED.
- On OWC research activities there is a European research project action IC1101 OPTICWISE of the COST Programme (European Cooperation in Science and Technology) funded by the European Science Foundation, allowing the coordination of nationally funded research on a European level. The Action aims to serve as a high-profile consolidated European scientific platform for interdisciplinary optical wireless communication (OWC) research activities. It was launched in November 2011 and will run until November 2015. More than 20 countries represented.

- The consumer and industry OWC technologies adoption is represented by the Li-Fi Consortium, founded in 2011 is a Non-profit organization, devoted to introduce optical wireless technology. Promotes the adoption of Light Fidelity (Li-Fi) products.
- An example of Asian awareness about OWC is the VLCC visible light communication consortium in Japan, established at 2007 in order to realize safe, ubiquitous telecommunication system using visible light through the activities of market research, promotion, and standardization.
- In the USA there are several OWC initiatives, including the "Smart Lighting Engineering Research Center", founded in 2008 by the National Science Foundation (NSF) is a partnership of Rensselaer Polytechnic Institute (lead institution), Boston University and the University of New Mexico. Outreach partners are Howard University, Morgan State University, and Rose-Hulman Institute of Technology.

Optical Communication

Optical communication, also known as optical telecommunication, is communication at a distance using light to carry information. It can be performed visually or by using electronic devices. The earliest basic forms of optical communication date back several millennia, while the earliest electrical device created to do so was the photophone, invented in 1880.

An optical communication system uses a transmitter, which encodes a message into an optical signal, a channel, which carries the signal to its destination, and a receiver, which reproduces the message from the received optical signal. When electronic equipment is not employed the 'receiver' is a person visually observing and interpreting a signal, which may be either simple (such as the presence of a beacon fire) or complex (such as lights using color codes or flashed in a Morse code sequence).

Free-space optical communication has been deployed in space, while terrestrial forms are naturally limited by geography, weather and the availability of light. This article provides a basic introduction to different forms of optical communication.

Forms

Visual techniques such as smoke signals, beacon fires, hydraulic telegraphs, ship flags and semaphore lines were the earliest forms of optical communication. Hydraulic telegraph semaphores date back to the 4th century BCE Greece. Distress flares are still used by mariners in emergencies, while lighthouses and navigation lights are used to communicate navigation hazards.

The heliograph uses a mirror to reflect sunlight to a distant observer. When a signaler tilts the mirror to reflect sunlight, the distant observer sees flashes of light that can be used to transmit a prearranged signaling code. Naval ships often use signal lamps and Morse code in a similar way.

Aircraft pilots often use visual approach slope indicator (VASI) projected light systems to land safely, especially at night. Military aircraft landing on an aircraft carrier use a similar system to land correctly on a carrier deck. The coloured light system communicates the aircraft's height relative to a standard landing glideslope. As well, airport control towers still use Aldis lamps to transmit instructions to aircraft whose radios have failed.

In the present day a variety of electronic systems optically transmit and receive information carried by pulses of light. Fiber-optic communication cables are now employed to send the great majority of the electronic data and long distance telephone calls that are not conveyed by either radio, terrestrial microwave or satellite. Free-space optical communications are also used every day in various applications.

Semaphore Line

A replica of one of Chappe's semaphore towers (18th century).

A 'semaphore telegraph', also called a 'semaphore line', 'optical telegraph', 'shutter telegraph chain', 'Chappe telegraph', or 'Napoleonic semaphore', is a system used for conveying information by means of visual signals, using towers with pivoting arms or shutters, also known as blades or paddles. Information is encoded by the position of the mechanical elements; it is read when the shutter is in a fixed position.

Semaphore lines were a precursor of the electrical telegraph. They were far faster than post riders for conveying a message over long distances, but far more expensive and less private than the electrical telegraph lines which would later replace them. The maximum distance that a pair of semaphore telegraph stations can bridge is limited by geography, weather and the availability of light; thus, in practical use, most optical telegraphs used lines of relay stations to bridge longer distances. Each relay station would also require its complement of skilled operator-observers to convey messages back and forth across the line.

The modern design of semaphores was first foreseen by the British polymath Robert Hooke, who first gave a vivid and comprehensive outline of visual telegraphy in an 1684 submission to the Royal Society. His proposal (which was motivated by military concerns following the Battle of Vienna the preceding year) was not put into practice during his lifetime.

The first operational optical semaphore line arrived in 1792, created by the French engineer Claude Chappe and his brothers, who succeeded in covering France with a network of 556 stations stretching a total distance of 4,800 kilometres (3,000 mi). It was used for military and national communications until the 1850s.

Many national services adopted signaling systems different from the Chappe system. For example, Britain and Sweden adopted systems of shuttered panels (in contradiction to the Chappe brothers' contention that angled rods are more visible). In Spain, the engineer Agustín de Betancourt developed his own system which was adopted by that state. This system was considered by many experts in Europe better than Chappe's, even in France.

These systems were popular in the late 18th to early 19th century but could not compete with the electrical telegraph, and went completely out of service by 1880.

Semaphore Signal Flags

A naval signaler transmitting a message by flag semaphore (2002).

Semaphore Flags is the system for conveying information at a distance by means of visual signals with hand-held flags, rods, disks, paddles, or occasionally bare or gloved hands. Information is encoded by the position of the flags, objects or arms; it is read when they are in a fixed position.

Semaphores were adopted and widely used (with hand-held flags replacing the mechanical arms of shutter semaphores) in the maritime world in the 19th century. They are still used during underway replenishment at sea and are acceptable for emergency communication in daylight or, using lighted wands instead of flags, at night.

The newer flag semaphore system uses two short poles with square flags, which a signaler holds in different positions to convey letters of the alphabet and numbers. The transmitter holds one pole in each hand, and extends each arm in one of eight possible directions. Except for in the rest position, the flags cannot overlap. The flags are colored differently based on whether the signals are sent by sea or by land. At sea, the flags are colored red and yellow (the Oscar flags), while on land, they are white and blue (the Papa flags). Flags are not required, they just make the characters more obvious.

Optical Fiber

Optical fiber is the most common type of channel for optical communications. The transmitters in optical fiber links are generally light-emitting diodes (LEDs) or laser diodes. Infrared light, rather than visible light is used more commonly, because optical fibers transmit infrared wavelengths with less attenuation and dispersion. The signal encoding is typically simple intensity modulation, although historically optical phase and frequency modulation have been demonstrated in the lab. The need for periodic signal regeneration was largely superseded by the introduction of the erbium-doped fiber amplifier, which extended link distances at significantly lower cost.

Signal Lamps

An air traffic controller holding a signal light gun that can be used to direct aircraft experiencing a radio failure (2007).

Signal lamps (such as Aldis lamps), are visual signaling devices for optical communication (typically using Morse code). Modern signal lamps are a focused lamp which can produce a pulse of light. In large versions this pulse is achieved by opening and closing shutters mounted in front of the lamp, either via a manually operated pressure switch or, in later versions, automatically.

With hand held lamps, a concave mirror is tilted by a trigger to focus the light into pulses. The lamps are usually equipped with some form of optical sight, and are most commonly deployed on naval vessels and also used in airport control towers with coded aviation light signals.

Aviation light signals are used in the case of a radio failure, an aircraft not equipped with a radio, or in the case of a hearing-impaired pilot. Air traffic controlers have long used signal light guns to direct such aircraft. The light gun's lamp has a focused bright beam capable of emitting three different colors: red, white and green. These colors may be flashing or steady, and provide different instructions to aircraft in flight or on the ground (for example, "cleared to land" or "cleared for takeoff"). Pilots can acknowledge the instructions by wiggling their plane's wings, moving their ailerons if they are on the ground, or by flashing their landing or navigation lights during night time. Only 12 simple standardized instructions are directed at aircraft using signal light guns as the system is not utilized with Morse code.

Photophone

The photophone (originally given an alternate name, radiophone) is a communication device which allowed for the transmission of speech on a beam of light. It was invented jointly by Alexander Graham Bell and his assistant Charles Sumner Tainter on February 19, 1880, at Bell's 1325 'L' Street laboratory in Washington, D.C. Both were later to become full associates in the Volta Laboratory Association, created and financed by Bell.

On June 21, 1880, Bell's assistant transmitted a wireless voice telephone message of considerable distance, from the roof of the Franklin School to the window of Bell's laboratory, some 213 meters (about 700 ft.) away.

Bell believed the photophone was his most important invention. Of the 18 patents granted in Bell's name alone, and the 12 he shared with his collaborators, four were for the photophone, which Bell referred to as his *'greatest achievement'*, telling a reporter shortly before his death that the photophone was *"the greatest invention [I have] ever made, greater than the telephone"*.

The photophone was a precursor to the fiber-optic communication systems which achieved popular worldwide usage starting in the 1980s. The master patent for the photophone (U.S. Patent 235,199 *Apparatus for Signalling and Communicating, called Photophone*), was issued in December 1880, many decades before its principles came to have practical applications.

Free-space Optical Communication

Free-space optics (FSO) systems are generally employed for 'last mile' telecommunications and can function over distances of several kilometers as long as there is a clear line of sight between the source and the destination, and the optical receiver can reliably decode the transmitted information. Other free-space systems can provide high-data-rate, long-range links using small, low-mass, low-power-consumption subsystems.

More generally, transmission of unguided optical signals is known as optical wireless communications (OWC). Examples include medium-range visible light communication and short-distance IrDA, using infrared LEDs.

Heliograph

Heliograph: Australians using a heliograph in North Africa (1940).

A heliograph is a wireless solar telegraph that signals by flashes of sunlight (generally using Morse code) reflected by a mirror. The flashes are produced by momentarily pivoting the mirror, or by interrupting the beam with a shutter.

The heliograph was a simple but effective instrument for instantaneous optical communication over long distances during the late 19th and early 20th century. Its main uses were in military, surveys and forest protection work. They were standard issue in the British and Australian armies until the 1960s, and were used by the Pakistani army as late as 1975.

Visible Light Communication

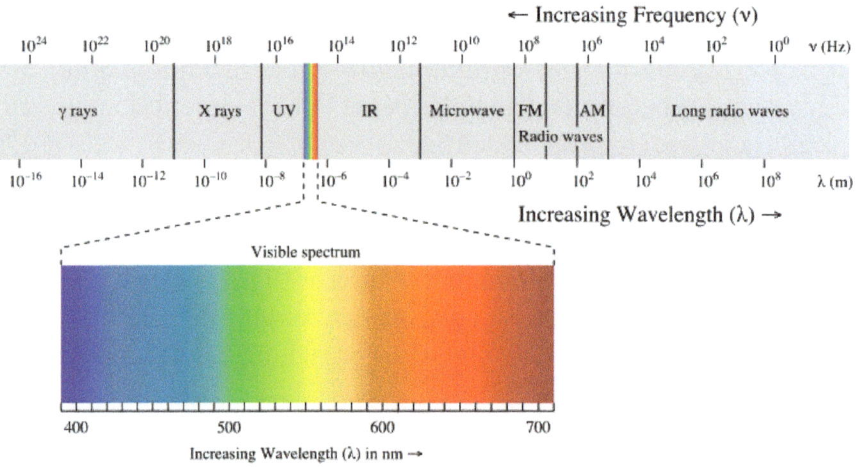

Visible light is only a small portion of the electromagnetic spectrum.

Visible light communication (VLC) is a data communications variant which uses visible light between 400 and 800 THz (780–375 nm). VLC is a subset of optical wireless communications technologies.

The technology uses fluorescent lamps (ordinary lamps, not special communications devices) to transmit signals at 10 kbit/s, or LEDs for up to 500 Mbit/s. Low rate data transmissions at 1 and 2 kilometres (0.6 and 1.2 mi) were demonstrated. RONJA achieves full Ethernet speed (10 Mbit/s) over the same distance thanks to larger optics and more powerful LEDs.

Specially designed electronic devices generally containing a photodiode receive signals from light sources, although in some cases a cell phone camera or a digital camera will be sufficient. The image sensor used in these devices is in fact an array of photodiodes (pixels) and in some applications its use may be preferred over a single photodiode. Such a sensor may provide either multi-channel communication (down to 1 pixel = 1 channel) or a spatial awareness of multiple light sources.

VLC can be used as a communications medium for ubiquitous computing, because light-producing devices (such as indoor/outdoor lamps, TVs, traffic signs, commercial displays and car headlights/taillights) are used everywhere. Using visible light is also less dangerous for high-power applications because humans can perceive it and act to protect their eyes from damage.

History

The history of Visible Light Communications (VLC) dates back to the 1880s in Washington, D.C. when the Scottish-born scientist Alexander Graham Bell invented the photophone, which transmitted speech on modulated sunlight over several hundred meters. This pre-dates the transmission of speech by radio.

More recent work began in 2003 at Nakagawa Laboratory, in Keio University, Japan, using LEDs to transmit data by visible light. A prototype of VLC had been presented by three undergraduate students at Universidad de Buenos Aires in 1995, resorting to the amplitude modulation of a 532 nm laser diode of 5 mW and photodiodes detector. Since then there have been numerous research activities focussed on VLC, notably by Smart Lighting Engineering Centre, Omega Project, COWA, ByteLight, Inc.,D-Light Project, UC-Light Centre, and work at Oxford University.

In 2006, researchers from CICTR at Penn State proposed a combination of power line communication (PLC) and white light LED to provide broadband access for indoor applications. This research suggested that VLC could be deployed as a perfect last-mile solution in the future.

In January 2010 a team of researchers from Siemens and Fraunhofer Institute for Telecommunications, Heinrich Hertz Institute in Berlin demonstrated transmission at 500 Mbit/s with a white LED over a distance of 5 metres (16 ft), and 100 Mbit/s over longer distance using five LEDs.

The VLC standardization process is conducted within IEEE Wireless Personal Area Networks working group (802.15).

In December 2010 St. Cloud, Minnesota, signed a contract with LVX Minnesota and became the first to commercially deploy this technology.

In July 2011 a live demonstration of high-definition video being transmitted from a standard LED lamp was shown at TED Global.

Recently, VLC-based indoor positioning system has become an attractive topic. ABI research forecasts that it could be a key solution to unlocking the $5 billion “indoor location market”. Publications have been coming from Nakagawa Laboratory, COWA at Penn State and other researchers around the world.

Another recent application is in the world of toys, thanks to cost-efficient and low-complexity implementation, which only requires one microcontroller and one LED as optical front-end.

VLCs can be used for providing security. They are especially useful in body sensor networks and personal area networks.

Recently Organic LEDs (OLED) have been used as optical transceivers to build up VLC communication links up to 10 Mbit/s.

In October 2014, Axrtek launched a commercial bidirectional RGB LED VLC system called MOMO that transmits down and up at speeds of 300 Mbit/s and with a range of 25 feet.

In May 2015, Philips collaborated with supermarket giant Carrefour to deliver VLC location-based services to shoppers' smartphones in a hypermarket in Lille, France. Indoor positioning systems based on VLC can be used in places such as hospitals, eldercare homes, warehouses, and large, open offices to locate people and control indoor robotic vehicles.

Heliograph

Fig. 1: Signaling with a Mance heliograph, 1910

A heliograph is a wireless solar telegraph that signals by flashes of sunlight (generally using Morse code) reflected by a mirror. The flashes are produced by momentarily pivoting the mirror, or by interrupting the beam with a shutter. The heliograph was a simple but effective instrument for instantaneous optical communication over long distances during the late 19th and early 20th century. Its main uses were military, survey and forest protection work. Heliographs were standard issue in the British and Australian armies until the 1960s, and were used by the Pakistani army as late as 1975.

Description

There were many heliograph types. Most heliographs were variants of the British Army Mance Mark V version (Fig.1). It used a mirror with a small unsilvered spot in the centre. The sender aligned the heliograph to the target by looking at the reflected target in the mirror and moving his head until the target was hidden by the unsilvered spot. Keeping his head still, he then adjusted the aiming rod so its cross wires bisected the target. He then turned up the sighting vane, which covered the cross wires with a diagram of a cross, and aligned the mirror with the tangent and elevation screws so the small shadow that was the reflection of the unsilvered spot hole was on the cross target. This indicated that the sunbeam was pointing at the target. The flashes were produced by a keying mechanism that tilted the mirror up a few degrees at the push of a lever at the back of the instrument. If the sun was in front of the sender, its rays were reflected directly from this mirror to the receiving station. If the sun was behind the sender, the sighting rod was replaced by a second mirror, to capture the sunlight from the main mirror and reflect it to the receiving station. The U.

S. Signal Corps heliograph mirror did not tilt. This type produced flashes by a shutter mounted on a second tripod (Fig 4).

Fig. 2: German Heliograph made by R. Fuess in Berlin (on display at the Museum of Communication in Frankfurt)

The heliograph had some great advantages. It allowed long distance communication without a fixed infrastructure, though it could also be linked to make a fixed network extending for hundreds of miles, as in the fort-to-fort network used for the Geronimo campaign. It was very portable, did not require any power source, and was relatively secure since it was invisible to those not near the axis of operation, and the beam was very narrow, spreading only 50 feet per mile of range. However, anyone in the beam with the correct knowledge could intercept signals without being detected. In the Boer War, where both sides used heliographs, tubes were sometimes used to decrease the dispersion of the beam. In some other circumstances, though, a narrow beam made it difficult to stay aligned with a moving target, as when communicating from shore to a moving ship, so the British issued a dispersing lens to broaden the heliograph beam from its natural diameter of 0.5 degrees to 15 degrees

The distance that heliograph signals could be seen depended on the clarity of the sky and the size of the mirrors used. A clear line of sight was required, and since the Earth's surface is curved, the highest convenient points were used. Under ordinary conditions, a flash could be seen 30 miles (48 km) with the naked eye, and much farther with a telescope. The maximum range was considered to be 10 miles for each inch of mirror diameter. Mirrors ranged from 1.5 inches to 12 inches or more. The record distance was established by a detachment of U.S. signal sergeants by the inter-operation of stations on Mount Ellen, Utah, and Mount Uncompahgre, Colorado, 183 miles (295 km) apart on September 17, 1894, with Signal Corps heliographs carrying mirrors only 8 inches square.

History

The German professor Carl Friedrich Gauss of the University of Göttingen developed and used a predecessor of the heliograph (the heliotrope) in 1821. His device directed a controlled beam of sunlight to a distant station to be used as a marker for geodetic survey work, and was suggested as a means of telegraphic communications. This is the first reliably documented heliographic device, despite much speculation about possible ancient incidents of sun-flash signalling, and the documented existence of other forms of ancient optical telegraphy.

Fig. 3 Ottoman heliograph crew at Huj during World War I, 1917

For example, one author in 1919 chose to "hazard the theory" that the mainland signals Roman emperor Tiberius watched for from Capri were mirror flashes, but admitted "there are no references in ancient writings to the use of signaling by mirrors", and that the documented means of ancient long-range visual telecommunications was by beacon fires and beacon smoke, not mirrors.

Similarly, the story that a shield was used as a heliograph at the Battle of Marathon is a modern myth, originating in the 1800s. Herodotus never mentioned any flash. What Herodotus did write was that someone was accused of having arranged to "hold up a shield as a signal". Suspicion grew in the 1900s that the flash theory was implausible. The conclusion after testing the theory was "Nobody flashed a shield at the Battle of Marathon".

In a letter dated 3 June 1778, John Norris, High Sheriff of Buckinghamshire, England, notes: "Did this day heliograph intelligence from Dr [Benjamin] Franklin in Paris to Wycombe". However there is little evidence that "heliograph" here is other than a misspelling of "holograph". The term "heliograph" for solar telegraphy did not enter the English language until the 1870s - even the word "telegraphy" was not coined until the 1790s.

Henry Christopher Mance (1840–1926), of the British Government Persian Gulf Telegraph Department, developed the first widely accepted heliograph about 1869 while stationed at Karachi, in the Bombay Presidency in British India. Mance was familiar with heliotropes by their use for the Great India Survey. The Mance Heliograph was operated easily by one man, and since it weighed about seven pounds, the operator could readily carry the device and its tripod. The British Army tested the heliograph in India at a range of 35 miles with favorable results. During the Jowaki Afridi expedition sent by the British-Indian government in 1877, the heliograph was first tested in war.

The simple and effective instrument that Mance invented was to be an important part of military communications for more than 60 years. The usefulness of heliographs was limited to daytimes with strong sunlight, but they were the most powerful type of visual signalling device known. In pre-radio times heliography was often the only means of communication that could span ranges of as much as 100 miles with a lightweight portable instrument.

Fig. 4: US Signal Service Heliograph, 1898

In the United States military, by January 1880, Colonel Nelson A. Miles had established a line of heliographs connecting Fort Keogh and Fort Custer, Montana, a distance of 140 miles. In 1890, Major W. J. Volkmar of the US Army, demonstrated in Arizona and New Mexico the possibility of performing communication by heliograph over a heliograph network aggregating 2,000 miles in length. The network of communication begun by General Miles in 1886, and continued by Lieutenant W. A. Glassford, was perfected in 1889 at ranges of 85, 88, 95, and 125 miles over a rugged and broken country, which was the stronghold of the Apache and other hostile Indian tribes.

By 1887, heliographs in use included not only the British Mance and Begbie heliographs, but also the American Grugan, Garner and Pursell heliographs. The Grugan and Pursell heliographs used shutters, and the others used movable mirrors operated by a finger key. The Mance, Grugam and Pursell heliographs used two tripods, and the others one. The signals could either be momentary flashes, or momentary obscurations. In 1888, the US Signal Service reviewed all of these devices, as well as the Finley Helio-Telegraph, and finding none completely suitable, developed the US Signal Service heliograph, a two-tripod, shutter-based machine of 13 7/8 lb. total weight, and ordered 100 for a total cost of $4,205. In 1893, the number of heliographs manufactured for the US Signal Service was 133.

The heyday of the heliograph was probably the Second Boer War in South Africa, where it was much used by both the British and the Boers. The terrain and climate, as well as the nature of the campaign, made heliography a logical choice. For night communications, the British used some large Aldis lamps, brought inland on railroad cars, and equipped with leaf-type shutters for keying a beam of light into dots and dashes. During the early stages of the war, the British garrisons were besieged in Kimberley, Ladysmith, and Mafeking. With land telegraph lines cut, the only contact with the outside world was via light-beam communication, helio by day, and Aldis lamps at night.

In 1909, the use of heliography for forestry protection was introduced in the United States. By 1920 such use was widespread in the US and beginning in Canada, and the heliograph was regarded as "next to the telephone, the most useful communication device that is at present available for forest-protection services". D.P. Godwin of the US Forestry Service invented a very portable (4.5 lb) heliograph of the single-tripod, shutter plus mirror type for forestry use.

Immediately prior to the outbreak of World War I, the cavalry regiments of the Russian Imperial Army were still being trained in heliograph communications to augment the efficiency of their scouting and reporting roles. The Red Army during the Russian Civil War made use of a series of heliograph stations to disseminate intelligence efficiently about basmachi rebel movements in Turkestan in 1926.

During World War II, South African and Australian forces used the heliograph against German forces in Libya and Egypt in 1941 and 1942.

The heliograph remained standard equipment for military signallers in the Australian and British armies until the 1960s, where it was considered a "low probability of intercept" type of communication. The Canadian Army was the last major army to have the heliograph as an issue item. By the time the mirror instruments were retired, they were seldom used for signalling. However, as recently as the 1980s, heliographs were used by Afghan forces during the Soviet invasion of Afghanistan. Signal mirrors are still included in survival kits for emergency signaling to search and rescue aircraft.

Automated Heliographs

Most heliographs of the 19th and 20th century were completely manual. The steps of aligning the heliograph on the target, co-aligning the reflected sunbeam with the heliograph, maintaining the sunbeam alignment as the sun moved, transcribing the message into flashes, modulating the sunbeam into those flashes, detecting the flashes at the receiving end, and transcribing the flashes into the message, were all manual steps. One notable exception – many French heliographs used clockwork heliostats to automatically steer out the sun's motion. By 1884, all active units of the "Mangin apparatus" (a dual-mode French military field optical telegraph that could use either lantern or sunlight) were equipped with clockwork heliostats. The Mangin apparatus with heliostat was still in service in 1917. Proposals to automate both the modulation of the sunbeam (by clockwork) and the detection (by electrical selenium photodetectors, or photographic means) date back to at least 1882. In 1961, the US Air Force was working on a space heliograph to signal between satellites

In May 2012, "Solar Beacon" robotic mirrors designed at UC Berkeley were mounted on the towers of the Golden Gate bridge, and a web site set up where the public could schedule times for the mirrors to signal with sun-flashes, entering the time and their latitude, longitude and altitude. The solar beacons were later moved to Sather Tower at UC Berkeley. By June 2012, the public could specify a "custom show" of up to 32 "on" or "off" periods of 4 seconds each, permitting the transmission of a few characters of Morse Code. The designer described the Solar Beacon as a "heliostat", not a "heliograph".

The first digitally controlled heliograph was designed and built in 2015. It was a semi-finalist in the Broadcom MASTERS competition.

Free-space Optical Communication

Free-space optical communication (FSO) is an optical communication technology that uses light propagating in free space to wirelessly transmit data for telecommunications or computer networking. "Free space" means air, outer space, vacuum, or something similar. This contrasts with using solids such as optical fiber cable or an optical transmission line. The technology is useful where the physical connections are impractical due to high costs or other considerations.

An 8-beam free space optics laser link, rated for 1 Gbit/s. The receptor is the large disc in the middle, the transmitters the smaller ones. At the top right corner is a monocular for assisting the alignment of the two heads.

History

A photophone receiver and headset, one half of Bell and Tainter's optical telecommunication system of 1880

Optical communications, in various forms, have been used for thousands of years. The Ancient Greeks used a coded alphabetic system of signalling with torches developed by Cleoxenus, Democleitus and Polybius. In the modern era, semaphores and wireless solar telegraphs called heliographs were developed, using coded signals to communicate with their recipients.

In 1880, Alexander Graham Bell and his assistant Charles Sumner Tainter created the photophone, at Bell's newly established Volta Laboratory in Washington, DC. Bell considered it his most important invention. The device allowed for the transmission of sound on a beam of light. On June 3, 1880, Bell conducted the world's first wireless telephone transmission between two buildings, some 213 meters (700 feet) apart.

Its first practical use came in military communication systems many decades later, first for optical telegraphy. German colonial troops used heliograph telegraphy transmitters during the Herero and Namaqua genocide starting in 1904, in German South-West Africa (today's Namibia) as did British, French, US or Ottoman signals.

WW I German Blinkgerät

During the trench warfare of World War I when wire communications were often cut, German signals used three types of optical Morse transmitters called *Blinkgerät*, the intermediate type for distances of up to 4 km (2.5 miles) at daylight and of up to 8 km (5 miles) at night, using red filters for undetected communications. Optical telephone communications were tested at the end of the war, but not introduced at troop level. In addition, special blinkgeräts were used for communication with airplanes, balloons, and tanks, with varying success.

A major technological step was to replace the Morse code by modulating optical waves in speech transmission. Carl Zeiss, Jena developed the *Lichtsprechgerät 80/80* (literal translation: optical speaking device) that the German army used in their World War II anti-aircraft defense units, or in bunkers at the Atlantic Wall.

The invention of lasers in the 1960s, revolutionized free space optics. Military organizations were particularly interested and boosted their development. However the technology lost market momentum when the installation of optical fiber networks for civilian uses was at its peak.

Many simple and inexpensive consumer remote controls use low-speed communication using infrared (IR) light. This is known as consumer IR technologies.

A recently declassified 1987 Pentagon report, reveals free-space lasers have been mounted on Israeli F-15 fighter jets for the purposes of surveillance, missile-tracking, and targeted weaponry.

Usage and Technologies

Free-space point-to-point optical links can be implemented using infrared laser light, although low-data-rate communication over short distances is possible using LEDs. Infrared Data Association (IrDA) technology is a very simple form of free-space optical communications. On the communications side the FSO technology is considered as a part of the Optical Wireless Communications applications. Free-space optics can be used for communications between spacecraft.

Current Market Demands

The demand for a high-speed (10+ Gbit/s) and long range (3–5 km) FSO system is apparent in the market place.

- In 2008, MRV Communications introduced a free-space optics (FSO)-based system with a data rate of 10 Gbit/s initially claiming a distance of 2 km at high availability. This equipment is no longer available; before end-of-life, the product's useful distance was changed down to 350 m.
- In 2013, the company MOSTCOM started to serially produce a new wireless communication system that also had a data rate of 10 Gbit/s as well as an improved range of up to 2.5 km, but to get to 99.99% uptime the designers used an RF hybrid solution, meaning the data rate drops to extremely low levels during atmospheric disturbances (typically down to 10 Mbit/s). In April 2014, the company with Scientific and Technological Centre "Fiord" demonstrated the transmission speed 30 Gbit/s under "laboratory conditions".
- LightPointe offers many similar hybrid solutions to MOSTCOM's offering.

Useful distances

The reliability of FSO units has always been a problem for commercial telecommunications. Consistently, studies find too many dropped packets and signal errors over small ranges (400 to 500 meters). This is from both independent studies, such as in the Czech republic, as well as formal internal nationwide studies, such as one conducted by MRV FSO staff. Military based studies consistently produce longer estimates for reliability, projecting the maximum range for terrestrial links is of the order of 2 to 3 km (1.2 to 1.9 mi). All studies agree the stability and quality of the link is highly dependent on atmospheric factors such as rain, fog, dust and heat.

Extending the Useful Distance

DARPA ORCA Official Concept Art created c. 2008

The main reason terrestrial communications have been limited to non-commercial telecommunications functions is fog. Fog consistently keeps FSO laser links over 500 meters from achieving a year-round bit error rate of 99.999%. Several entities are continually attempting to overcome these key disadvantages to FSO communications and field a system with a better quality of service. DARPA has sponsored over US$130 million in research towards this effort, with the ORCA and ORCLE programs.

Other non-government groups are fielding tests to evaluate different technologies that some claim have the ability to address key FSO adoption challenges. As of October 2014, none have fielded a working system that addresses the most common atmospheric events.

FSO research from 1998–2006 in the private sector totaled $407.1 million, divided primarily among four start-up companies. All four failed to deliver products that would meet telecommunications quality and distance standards:

- Terabeam received approximately $226 million in funding. AT&T and Lucent backed this attempt. The work ultimately failed, and the company reorganized in 2004.
- AirFiber received $96.1 million in funding, and never solved the weather issue. They sold out to MRV communications in 2003, and MRV sold their FSO units until 2012 when the end-of-life was abruptly announced for the Terescope series.
- LightPointe Communications received $76 million in start-up funds, and eventually reorganized to sell hybrid FSO-RF units to overcome the weather-based challenges.
- The Maxima Corporation published its operating theory in Science (magazine), and received $9 million in funding before permanently shutting down. No known spin-off or purchase followed this effort.
- Wireless Excellence developed and launched CableFree UNITY solutions that combine FSO with Millimeter Wave and Radio technologies to extend distance, capacity and availability, with a goal of making FSO a more useful and practical technology.

One private company published a paper on November 20, 2014, claiming they had achieved commercial reliability (99.999% availability) in extreme fog. There is no indication this product is currently commercially available.

Extraterrestrial

The massive advantages of laser communication in space have multiple space agencies racing to develop a stable space communication platform, with many significant demonstrations and achievements. As of 18 December 2014, *no laser communication system is in use in space.*

Demonstrations In Space:

The first gigabit laser-based communication was achieved by the European Space Agency and called the European Data Relay System (EDRS) on November 28, 2014. The initial images have just been demonstrated, and a working system is expected to be in place in the 2015–2016 time frame.

NASA's OPALS announced a breakthrough in space-to-ground communication December 9, 2014, uploading 175 megabytes in 3.5 seconds. Their system is also able to re-acquire tracking after the signal was lost due to cloud cover.

In January 2013, NASA used lasers to beam an image of the Mona Lisa to the Lunar Reconnaissance Orbiter roughly 390,000 km (240,000 mi) away. To compensate for atmospheric interference, an error correction code algorithm similar to that used in CDs was implemented.

A two-way distance record for communication was set by the Mercury laser altimeter instrument aboard the MESSENGER spacecraft, and was able to communicate across a distance of 24 million km (15 million miles), as the craft neared Earth on a fly-by in May, 2005. The previous record had been set with a one-way detection of laser light from Earth, by the Galileo probe, of 6 million km in 1992. Quote from Laser Communication in Space Demonstrations (EDRS)

LEDs

RONJA is a free implementation of FSO using high-intensity LEDs.

In 2001, Twibright Labs released Ronja Metropolis, an open source DIY 10 Mbit/s full duplex LED FSO over 1.4 km In 2004, a Visible Light Communication Consortium was formed in Japan. This was based on work from researchers that used a white LED-based space lighting system for indoor local area network (LAN) communications. These systems present advantages over traditional UHF RF-based systems from improved isolation between systems, the size and cost of receivers/transmitters, RF licensing laws and by combining space lighting and communication into the same system. In January 2009, a task force for visible light communication was formed by the Institute of Electrical and Electronics Engineers working group for wireless personal area network standards known as IEEE 802.15.7. A trial was announced in 2010, in St. Cloud, Minnesota.

Amateur radio operators have achieved significantly farther distances using incoherent sources of light from high-intensity LEDs. One reported 173 miles (278 km) in 2007. However, physical limitations of the equipment used limited bandwidths to about 4 kHz. The high sensitivities required of the detector to cover such distances made the internal capacitance of the photodiode used a dominant factor in the high-impedance amplifier which followed it, thus naturally forming a low-pass filter with a cut-off frequency in the 4 kHz range. From the other side use of lasers radiation source allows to reach very high data rates which are comparable to fiber communications.

Projected data rates and future data rate claims vary. A low-cost white LED (GaN-phosphor) which could be used for space lighting can typically be modulated up to 20 MHz. Data rates of over 100 Mbit/s can be easily achieved using efficient modulation schemes and Siemens claimed to have achieved over 500 Mbit/s in 2010. Research published in 2009, used a similar system for traffic control of automated vehicles with LED traffic lights.

In September 2013, pureLiFi, the Edinburgh start-up working on Li-Fi, also demonstrated high speed point-to-point connectivity using any off-the-shelf LED light bulb. In previous work, high bandwidth specialist LEDs have been used to achieve the high data rates. The new system, the Li-1st, maximizes the available optical bandwidth for any LED device, thereby reducing the cost and improving the performance of deploying indoor FSO systems.

Engineering Details

Typically, best use scenarios for this technology are:

- LAN-to-LAN connections on campuses at Fast Ethernet or Gigabit Ethernet speeds
- LAN-to-LAN connections in a city, a metropolitan area network
- To cross a public road or other barriers which the sender and receiver do not own
- Speedy service delivery of high-bandwidth access to optical fiber networks
- Converged Voice-Data-Connection
- Temporary network installation (for events or other purposes)
- Reestablish high-speed connection quickly (disaster recovery)
- As an alternative or upgrade add-on to existing wireless technologies
    - Especially powerful in combination with auto aiming systems, this way you could power moving cars or you can power your laptop while you move or use auto-aiming nodes to create a network with other nodes.
- As a safety add-on for important fiber connections (redundancy)
- For communications between spacecraft, including elements of a satellite constellation
- For inter- and intra-chip communication

The light beam can be very narrow, which makes FSO hard to intercept, improving security. In any case, it is comparatively easy to encrypt any data traveling across the FSO connection for additional security. FSO provides vastly improved electromagnetic interference (EMI) behavior compared to using microwaves.

Technical Advantages

- Ease of deployment
- Can be used to power devices

- License-free long-range operation (in contrast with radio communication)
- High bit rates
- Low bit error rates
- Immunity to electromagnetic interference
- Full duplex operation
- Protocol transparency
- Increased security when working with narrow beam(s)
- No Fresnel zone necessary
- Reference open source implementation

Range Limiting Factors

For terrestrial applications, the principal limiting factors are:

- Fog (10 to ~100 dB/km attenuation)
- Beam dispersion
- Atmospheric absorption
- Rain
- Snow
- Terrestrial scintillation
- Interference from background light sources (including the Sun)
- Shadowing
- Pointing stability in wind
- Pollution / smog

These factors cause an attenuated receiver signal and lead to higher bit error ratio (BER). To overcome these issues, vendors found some solutions, like multi-beam or multi-path architectures, which use more than one sender and more than one receiver. Some state-of-the-art devices also have larger fade margin (extra power, reserved for rain, smog, fog). To keep an eye-safe environment, good FSO systems have a limited laser power density and support laser classes 1 or 1M. Atmospheric and fog attenuation, which are exponential in nature, limit practical range of FSO devices to several kilometres.

References

- Holzmann, Gerard (1995). Björn Pehrson, ed. The early history of data networks. p. 10. ISBN 0818667826. Retrieved 18 November 2012.

- Dunnington, G. Waldo (2004). Carl Friedrich Gauss: Titan of Science. New York: The Mathematical Association of America. pp. 122–127. ISBN 0-88385-547-X.
- Woods, Daniel (2008). "Heliograph and Mirrors". In Sterling, Christopher. Military Communications: From Ancient Times to the 21st Century. ABC-CLIO. p. 208. ISBN 1851097325.
- Chapter 2: Semaphore Signalling ISBN 978-0-86341-327-8 Communications: an international history of the formative years R. W. Burns, 2004
- Sekunda, Nicholas (2002). Marathon 490 BC: The First Persian Invasion Of Greece. Osprey Publishing. p. 73. ISBN 1841760005.
- Deacon, Richard (1978). The Silent War: a History of Western Naval Intelligence. David & Charles. p. 21. ISBN 978-0715375570.
- Mary Kay Carson (2007). Alexander Graham Bell: Giving Voice To The World. Sterling Biographies. New York: Sterling Publishing. pp. 76–78. ISBN 978-1-4027-3230-0.
- Greely, Adolphus (August 1899). "The Evolution of the Signal Corps". Ainslee's Magazine. IV (1): 17. Retrieved 1 September 2014.
- Bruce V. Bigelow (June 16, 2006). "Zapped of its potential, Rooftop laser startups falter, but debate on high-speed data technology remains". Retrieved October 26, 2014.
- Fred Dawson (May 1, 2000). "TeraBeam, Lucent Extend Bandwidth Limits, Multichannel News, Vol 21 Issue 18 Pg 160". Retrieved October 27, 2014.

7

Wireless Security: A Comprehensive Study

Wireless security is very important in today's times. It prevents unapproved access to computers that use wireless networks. A brief explanation of intrusion detection system, RADIUS and Wi-Fi protected access is also provided, which provides a better understanding of wireless security.

Wireless Security

Wireless security is the prevention of unauthorized access or damage to computers using wireless networks. The most common types of wireless security are Wired Equivalent Privacy (WEP) and Wi-Fi Protected Access (WPA). WEP is a notoriously weak security standard. The password it uses can often be cracked in a few minutes with a basic laptop computer and widely available software tools. WEP is an old IEEE 802.11 standard from 1999, which was outdated in 2003 by WPA, or Wi-Fi Protected Access. WPA was a quick alternative to improve security over WEP. The current standard is WPA2; some hardware cannot support WPA2 without firmware upgrade or replacement. WPA2 uses an encryption device that encrypts the network with a 256-bit key; the longer key length improves security over WEP.

An example wireless router, that can implement **wireless security** features

Many laptop computers have wireless cards pre-installed. The ability to enter a network while mobile has great benefits. However, wireless networking is prone to some security issues. Hackers have found wireless networks relatively easy to break into, and even use wireless technology to hack into wired networks. As a result, it is very important that enterprises define effective wireless security policies that guard against unauthorized access to important resources. Wireless Intrusion Prevention Systems (WIPS) or Wireless Intrusion Detection Systems (WIDS) are commonly used to enforce wireless security policies.

Security settings panel for a DD-WRT router

The risks to users of wireless technology have increased as the service has become more popular. There were relatively few dangers when wireless technology was first introduced. Hackers had not yet had time to latch on to the new technology, and wireless networks were not commonly found in the work place. However, there are many security risks associated with the current wireless protocols and encryption methods, and in the carelessness and ignorance that exists at the user and corporate IT level. Hacking methods have become much more sophisticated and innovative with wireless access. Hacking has also become much easier and more accessible with easy-to-use Windows- or Linux-based tools being made available on the web at no charge.

Some organizations that have no wireless access points installed do not feel that they need to address wireless security concerns. In-Stat MDR and META Group have estimated that 95% of all corporate laptop computers that were planned to be purchased in 2005 were equipped with wireless cards. Issues can arise in a supposedly non-wireless organization when a wireless laptop is plugged into the corporate network. A hacker could sit out in the parking lot and gather information from it through laptops and/or other devices, or even break in through this wireless card–equipped laptop and gain access to the wired network.

Background

Anyone within the geographical network range of an open, unencrypted wireless network can "sniff", or capture and record, the traffic, gain unauthorized access to internal network resources as well as to the internet, and then use the information and resources to perform disruptive or illegal acts. Such security breaches have become important concerns for both enterprise and home networks.

If router security is not activated or if the owner deactivates it for convenience, it creates a free hotspot. Since most 21st-century laptop PCs have wireless networking built in, they don't need a third-party adapter such as a PCMCIA Card or USB dongle. Built-in wireless networking might be enabled by default, without the owner realizing it, thus broadcasting the laptop's accessibility to any computer nearby.

Modern operating systems such as Linux, macOS, or Microsoft Windows make it fairly easy to set up a PC as a wireless LAN "base station" using Internet Connection Sharing, thus allowing all the PCs in the home to access the Internet through the "base" PC. However, lack of knowledge among users about the security issues inherent in setting up such systems often may allow others nearby access to the connection. Such "piggybacking" is usually achieved without the wireless network operator's knowledge; it may even be without the knowledge of the intruding user if their computer automatically selects a nearby unsecured wireless network to use as an access point.

The Threat Situation

Wireless security is just an aspect of computer security; however, organizations may be particularly vulnerable to security breaches caused by rogue access points.

If an employee (trusted entity) brings in a wireless router and plugs it into an unsecured switchport, the entire network can be exposed to anyone within range of the signals. Similarly, if an employee adds a wireless interface to a networked computer using an open USB port, they may create a breach in network security that would allow access to confidential materials. However, there are effective countermeasures (like disabling open switchports during switch configuration and VLAN configuration to limit network access) that are available to protect both the network and the information it contains, but such countermeasures must be applied uniformly to all network devices.

Threats and Vulnerabilites in an Industrial (M2M) Context

Due to its availability and low cost, the use of wireless communication technologies increases in domains beyond the originally intended usage areas, e.g. M2M communication in industrial applications. Such industrial applications often have specific security requirements. Hence, it is important to understand the characteristics of such applications and evaluate the vulnerabilities bearing the highest risk in this context. Evaluation of these vulnerabilities and the resulting vulnerability catalogs in an industrial context when considering WLAN, NFC and ZigBee are available.

The Mobility Advantage

Wireless networks are very common, both for organizations and individuals. Many laptop computers have wireless cards pre-installed. The ability to enter a network while mobile has great benefits. However, wireless networking is prone to some security issues. Hackers have found wireless networks relatively easy to break into, and even use wireless technology to hack into wired networks. As a result, it is very important that enterprises define effective wireless security policies that guard against unauthorized access to important resources. Wireless Intrusion Prevention Systems (WIPS) or Wireless Intrusion Detection Systems (WIDS) are commonly used to enforce wireless security policies.

The Air Interface and Link Corruption Risk

There were relatively few dangers when wireless technology was first introduced, as the effort to maintain the communication was high and the effort to intrude is always higher. The variety of risks to users of wireless technology have increased as the service has become more popular and the technology more commonly available. Today there are a great number of security risks associated with the current wireless protocols and encryption methods, as carelessness and ignorance exists at the user and corporate IT level. Hacking methods have become much more sophisticated and innovative with wireless.

Modes of Unauthorized Access

The modes of unauthorised access to links, to functions and to data is as variable as the respective entities make use of program code. There does not exist a full scope model of such threat. To some extent the prevention relies on known modes and methods of attack and relevant methods for suppression of the applied methods. However, each new mode of operation will create new options of threatening. Hence prevention requires a steady drive for improvement. The described modes of attack are just a snapshot of typical methods and scenarios where to apply.

Accidental Association

Violation of the security perimeter of a corporate network can come from a number of different methods and intents. One of these methods is referred to as "accidental association". When a user turns on a computer and it latches on to a wireless access point from a neighboring company's overlapping network, the user may not even know that this has occurred. However, it is a security breach in that proprietary company information is exposed and now there could exist a link from one company to the other. This is especially true if the laptop is also hooked to a wired network.

Accidental association is a case of wireless vulnerability called as "mis-association". Mis-association can be accidental, deliberate (for example, done to bypass corporate firewall) or it can result from deliberate attempts on wireless clients to lure them into connecting to attacker's APs.

Malicious Association

"Malicious associations" are when wireless devices can be actively made by attackers to connect to a company network through their laptop instead of a company access point (AP). These types of laptops are known as "soft APs" and are created when a cyber criminal runs some software that makes his/her wireless network card look like a legitimate access point. Once the thief has gained access, he/she can steal passwords, launch attacks on the wired network, or plant trojans. Since wireless networks operate at the Layer 2 level, Layer 3 protections such as network authentication and virtual private networks (VPNs) offer no barrier. Wireless 802.1x authentications do help with some protection but are still vulnerable to hacking. The idea behind this type of attack may not be to break into a VPN or other security measures. Most likely the criminal is just trying to take over the client at the Layer 2 level.

Ad Hoc Networks

Ad hoc networks can pose a security threat. Ad hoc networks are defined as [peer to peer] networks between wireless computers that do not have an access point in between them. While these types of networks usually have little protection, encryption methods can be used to provide security.

The security hole provided by Ad hoc networking is not the Ad hoc network itself but the bridge it provides into other networks, usually in the corporate environment, and the unfortunate default settings in most versions of Microsoft Windows to have this feature turned on unless explicitly disabled. Thus the user may not even know they have an unsecured Ad hoc network in operation on their computer. If they are also using a wired or wireless infrastructure network at the same time, they are providing a bridge to the secured organizational network through the unsecured Ad hoc connection. Bridging is in two forms. A direct bridge, which requires the user actually configure a bridge between the two connections and is thus unlikely to be initiated unless explicitly desired, and an indirect bridge which is the shared resources on the user computer. The indirect bridge may expose private data that is shared from the user's computer to LAN connections, such as shared folders or private Network Attached Storage, making no distinction between authenticated or private connections and unauthenticated Ad-Hoc networks. This presents no threats not already familiar to open/public or unsecured wifi access points, but firewall rules may be circumvented in the case of poorly configured operating systems or local settings.

Non-traditional Networks

Non-traditional networks such as personal network Bluetooth devices are not safe from hacking and should be regarded as a security risk. Even barcode readers, handheld PDAs, and wireless printers and copiers should be secured. These non-traditional networks can be easily overlooked by IT personnel who have narrowly focused on laptops and access points.

Identity Theft (MAC Spoofing)

Identity theft (or MAC spoofing) occurs when a hacker is able to listen in on network traffic and identify the MAC address of a computer with network privileges. Most wireless systems allow some kind of MAC filtering to allow only authorized computers with specific MAC IDs to gain access and utilize the network. However, programs exist that have network "sniffing" capabilities. Combine these programs with other software that allow a computer to pretend it has any MAC address that the hacker desires, and the hacker can easily get around that hurdle.

MAC filtering is effective only for small residential (SOHO) networks, since it provides protection only when the wireless device is "off the air". Any 802.11 device "on the air" freely transmits its unencrypted MAC address in its 802.11 headers, and it requires no special equipment or software to detect it. Anyone with an 802.11 receiver (laptop and wireless adapter) and a freeware wireless packet analyzer can obtain the MAC address of any transmitting 802.11 within range. In an organizational environment, where most wireless devices are "on the air" throughout the active working shift, MAC filtering provides only a false sense of security since it prevents only "casual" or unintended connections to the organizational infrastructure and does nothing to prevent a directed attack.

Man-in-the-middle Attacks

A man-in-the-middle attacker entices computers to log in to a computer which is set up as a soft AP (Access Point). Once this is done, the hacker connects to a real access point through another wireless card offering a steady flow of traffic through the transparent hacking computer to the real network. The hacker can then sniff the traffic. One type of man-in-the-middle attack relies on security faults in challenge and handshake protocols to execute a "de-authentication attack". This attack forces AP-connected computers to drop their connections and reconnect with the hacker's soft AP (disconnects the user from the modem so they have to connect again using their password which one can extract from the recording of the event). Man-in-the-middle attacks are enhanced by software such as LANjack and AirJack which automate multiple steps of the process, meaning what once required some skill can now be done by script kiddies. Hotspots are particularly vulnerable to any attack since there is little to no security on these networks..

Denial of Service

A Denial-of-Service attack (DoS) occurs when an attacker continually bombards a targeted AP (Access Point) or network with bogus requests, premature successful connection messages, failure messages, and/or other commands. These cause legitimate users to not be able to get on the network and may even cause the network to crash. These attacks rely on the abuse of protocols such as the Extensible Authentication Protocol (EAP).

The DoS attack in itself does little to expose organizational data to a malicious attacker, since the interruption of the network prevents the flow of data and actually indirectly protects data by preventing it from being transmitted. The usual reason for performing a DoS attack is to observe the recovery of the wireless network, during which all of the initial handshake codes are re-transmitted by all devices, providing an opportunity for the malicious attacker to record these codes and use various cracking tools to analyze security weaknesses and exploit them to gain unauthorized access to the system. This works best on weakly encrypted systems such as WEP, where there are a number of tools available which can launch a dictionary style attack of "possibly accepted" security keys based on the "model" security key captured during the network recovery.

Network Injection

In a network injection attack, a hacker can make use of access points that are exposed to non-filtered network traffic, specifically broadcasting network traffic such as "Spanning Tree" (802.1D), OSPF, RIP, and HSRP. The hacker injects bogus networking re-configuration commands that affect routers, switches, and intelligent hubs. A whole network can be brought down in this manner and require rebooting or even reprogramming of all intelligent networking devices.

Caffe Latte Attack

The Caffe Latte attack is another way to defeat WEP. It is not necessary for the attacker to be in the area of the network using this exploit. By using a process that targets the Windows wireless stack, it is possible to obtain the WEP key from a remote client. By sending a flood of encrypted ARP requests, the assailant takes advantage of the shared key authentication and the message modification flaws in 802.11 WEP. The attacker uses the ARP responses to obtain the WEP key in less than 6 minutes.

Wireless Intrusion Prevention Concepts

There are three principal ways to secure a wireless network.

- For closed networks (like home users and organizations) the most common way is to configure access restrictions in the access points. Those restrictions may include encryption and checks on MAC address. Wireless Intrusion Prevention Systems can be used to provide wireless LAN security in this network model.

- For commercial providers, hotspots, and large organizations, the preferred solution is often to have an open and unencrypted, but completely isolated wireless network. The users will at first have no access to the Internet nor to any local network resources. Commercial providers usually forward all web traffic to a captive portal which provides for payment and/or authorization. Another solution is to require the users to connect securely to a privileged network using VPN.
- Wireless networks are less secure than wired ones; in many offices intruders can easily visit and hook up their own computer to the wired network without problems, gaining access to the network, and it is also often possible for remote intruders to gain access to the network through backdoors like Back Orifice. One general solution may be end-to-end encryption, with independent authentication on all resources that shouldn't be available to the public.

There is no ready designed system to prevent from fraudulent usage of wireless communication or to protect data and functions with wirelessly communicating computers and other entities. However, there is a system of qualifying the taken measures as a whole according to a common understanding what shall be seen as state of the art. The system of qualifying is an international consensus as specified in ISO/IEC 15408.

A Wireless Intrusion Prevention System

A Wireless Intrusion Prevention System (WIPS) is a concept for the most robust way to counteract wireless security risks. However such WIPS does not exist as a ready designed solution to implement as a software package. A WIPS is typically implemented as an overlay to an existing Wireless LAN infrastructure, although it may be deployed standalone to enforce no-wireless policies within an organization. WIPS is considered so important to wireless security that in July 2009, the Payment Card Industry Security Standards Council published wireless guidelines for PCI DSS recommending the use of WIPS to automate wireless scanning and protection for large organizations.

Security Measures

There are a range of wireless security measures, of varying effectiveness and practicality.

SSID Hiding

A simple but ineffective method to attempt to secure a wireless network is to hide the SSID (Service Set Identifier). This provides very little protection against anything but the most casual intrusion efforts.

MAC ID Filtering

One of the simplest techniques is to only allow access from known, pre-approved MAC addresses. Most wireless access points contain some type of MAC ID filtering. However, an attacker can simply sniff the MAC address of an authorized client and spoof this address.

Static IP Addressing

Typical wireless access points provide IP addresses to clients via DHCP. Requiring clients to set their own addresses makes it more difficult for a casual or unsophisticated intruder to log onto the network, but provides little protection against a sophisticated attacker.

802.11 Security

IEEE 802.1X is the IEEE Standard authentication mechanisms to devices wishing to attach to a Wireless LAN.

Regular WEP

The Wired Equivalent Privacy (WEP) encryption standard was the original encryption standard for wireless, but since 2004 with the ratification WPA2 the IEEE has declared it "deprecated", and while often supported, it is seldom or never the default on modern equipment.

Concerns were raised about its security as early as 2001, dramatically demonstrated in 2005 by the FBI, yet in 2007 T.J. Maxx admitted a massive security breach due in part to a reliance on WEP and the Payment Card Industry took until 2008 to prohibit its use - and even then allowed existing use to continue until June 2010.

WPAv1

The Wi-Fi Protected Access (WPA and WPA2) security protocols were later created to address the problems with WEP. If a weak password, such as a dictionary word or short character string is used, WPA and WPA2 can be cracked. Using a long enough random password (e.g. 14 random letters) or passphrase (e.g. 5 randomly chosen words) makes pre-shared key WPA virtually uncrackable. The second generation of the WPA security protocol (WPA2) is based on the final IEEE 802.11i amendment to the 802.11 standard and is eligible for FIPS 140-2 compliance. With all those encryption schemes, any client in the network that knows the keys can read all the traffic.

Wi-Fi Protected Access (WPA) is a software/firmware improvement over WEP. All regular WLAN-equipment that worked with WEP are able to be simply upgraded and no new equipment needs to be bought. WPA is a trimmed-down version of the 802.11i security standard that was developed by the IEEE 802.11 to replace WEP. The TKIP encryption algorithm was developed for WPA to provide improvements to WEP that could be fielded as firmware upgrades to existing 802.11 devices. The WPA profile also provides optional support for the AES-CCMP algorithm that is the preferred algorithm in 802.11i and WPA2.

WPA Enterprise provides RADIUS based authentication using 802.1x. WPA Personal uses a pre-shared Shared Key (PSK) to establish the security using an 8 to 63 character passphrase. The PSK may also be entered as a 64 character hexadecimal string. Weak PSK passphrases can be broken using off-line dictionary attacks by capturing the messages in the four-way exchange when the client reconnects after being deauthenticated. Wireless suites such as aircrack-ng can crack a weak passphrase in less than a minute. Other WEP/WPA crackers are AirSnort and Auditor Security Collection. Still, WPA Personal is secure when used with 'good' passphrases or a full 64-character hexadecimal key.

There was information, however, that Erik Tews (the man who created the fragmentation attack against WEP) was going to reveal a way of breaking the WPA TKIP implementation at Tokyo's PacSec security conference in November 2008, cracking the encryption on a packet in between

12–15 minutes. Still, the announcement of this 'crack' was somewhat overblown by the media, because as of August, 2009, the best attack on WPA (the Beck-Tews attack) is only partially successful in that it only works on short data packets, it cannot decipher the WPA key, and it requires very specific WPA implementations in order to work.

Additions to WPAv1

In addition to WPAv1, TKIP, WIDS and EAP may be added alongside. Also, VPN-networks (non-continuous secure network connections) may be set up under the 802.11-standard. VPN implementations include PPTP, L2TP, IPsec and SSH. However, this extra layer of security may also be cracked with tools such as Anger, Deceit and Ettercap for PPTP; and ike-scan, IKEProbe, ipsec-trace, and IKEcrack for IPsec-connections.

TKIP

This stands for Temporal Key Integrity Protocol and the acronym is pronounced as tee-kip. This is part of the IEEE 802.11i standard. TKIP implements per-packet key mixing with a re-keying system and also provides a message integrity check. These avoid the problems of WEP.

EAP

The WPA-improvement over the IEEE 802.1X standard already improved the authentication and authorization for access of wireless and wired LANs. In addition to this, extra measures such as the Extensible Authentication Protocol (EAP) have initiated an even greater amount of security. This, as EAP uses a central authentication server. Unfortunately, during 2002 a Maryland professor discovered some shortcomings. Over the next few years these shortcomings were addressed with the use of TLS and other enhancements. This new version of EAP is now called Extended EAP and is available in several versions; these include: EAP-MD5, PEAPv0, PEAPv1, EAP-MSCHAPv2, LEAP, EAP-FAST, EAP-TLS, EAP-TTLS, MSCHAPv2, and EAP-SIM.

EAP-versions

EAP-versions include LEAP, PEAP and other EAP's.

LEAP

This stands for the Lightweight Extensible Authentication Protocol. This protocol is based on 802.1X and helps minimize the original security flaws by using WEP and a sophisticated key management system. This EAP-version is safer than EAP-MD5. This also uses MAC address authentication. LEAP is not secure; THC-LeapCracker can be used to break Cisco's version of LEAP and be used against computers connected to an access point in the form of a dictionary attack. Anwrap and asleap finally are other crackers capable of breaking LEAP.

PEAP

This stands for Protected Extensible Authentication Protocol. This protocol allows for a secure transport of data, passwords, and encryption keys without the need of a certificate server. This was developed by Cisco, Microsoft, and RSA Security.

Other EAPs There are other types of Extensible Authentication Protocol implementations that are based on the EAP framework. The framework that was established supports existing EAP types as well as future authentication methods. EAP-TLS offers very good protection because of its mutual authentication. Both the client and the network are authenticated using certificates and per-session WEP keys. EAP-FAST also offers good protection. EAP-TTLS is another alternative made by Certicom and Funk Software. It is more convenient as one does not need to distribute certificates to users, yet offers slightly less protection than EAP-TLS.

Restricted Access Networks

Solutions include a newer system for authentication, IEEE 802.1x, that promises to enhance security on both wired and wireless networks. Wireless access points that incorporate technologies like these often also have routers built in, thus becoming wireless gateways.

End-to-end Encryption

One can argue that both layer 2 and layer 3 encryption methods are not good enough for protecting valuable data like passwords and personal emails. Those technologies add encryption only to parts of the communication path, still allowing people to spy on the traffic if they have gained access to the wired network somehow. The solution may be encryption and authorization in the application layer, using technologies like SSL, SSH, GnuPG, PGP and similar.

The disadvantage with the end-to-end method is, it may fail to cover all traffic. With encryption on the router level or VPN, a single switch encrypts all traffic, even UDP and DNS lookups. With end-to-end encryption on the other hand, each service to be secured must have its encryption "turned on", and often every connection must also be "turned on" separately. For sending emails, every recipient must support the encryption method, and must exchange keys correctly. For Web, not all web sites offer https, and even if they do, the browser sends out IP addresses in clear text.

The most prized resource is often access to Internet. An office LAN owner seeking to restrict such access will face the nontrivial enforcement task of having each user authenticate themselves for the router.

802.11i Security

The newest and most rigorous security to implement into WLAN's today is the 802.11i RSN-standard. This full-fledged 802.11i standard (which uses WPAv2) however does require the newest hardware (unlike WPAv1), thus potentially requiring the purchase of new equipment. This new hardware required may be either AES-WRAP (an early version of 802.11i) or the newer and better AES-CCMP-equipment. One should make sure one needs WRAP or CCMP-equipment, as the 2 hardware standards are not compatible.

WPAv2

WPA2 is a WiFi Alliance branded version of the final 802.11i standard. The primary enhancement over WPA is the inclusion of the AES-CCMP algorithm as a mandatory feature. Both WPA and WPA2 support EAP authentication methods using RADIUS servers and preshared key (PSK).

The number of WPA and WPA2 networks are increasing, while the number of WEP networks are decreasing, because of the security vulnerabilities in WEP.

WPA2 has been found to have at least one security vulnerability, nicknamed Hole196. The vulnerability uses the WPA2 Group Temporal Key (GTK), which is a shared key among all users of the same BSSID, to launch attacks on other users of the same BSSID. It is named after page 196 of the IEEE 802.11i specification, where the vulnerability is discussed. In order for this exploit to be performed, the GTK must be known by the attacker.

Additions to WPAv2

Unlike 802.1X, 802.11i already has most other additional security-services such as TKIP. Just as with WPAv1, WPAv2 may work in cooperation with EAP and a WIDS.

WAPI

This stands for WLAN Authentication and Privacy Infrastructure. This is a wireless security standard defined by the Chinese government.

Smart cards, USB Tokens, and Software Tokens

This is a very strong form of security. When combined with some server software, the hardware or software card or token will use its internal identity code combined with a user entered PIN to create a powerful algorithm that will very frequently generate a new encryption code. The server will be time synced to the card or token. This is a very secure way to conduct wireless transmissions. Companies in this area make USB tokens, software tokens, and smart cards. They even make hardware versions that double as an employee picture badge. Currently the safest security measures are the smart cards / USB tokens. However, these are expensive. The next safest methods are WPA2 or WPA with a RADIUS server. Any one of the three will provide a good base foundation for security. The third item on the list is to educate both employees and contractors on security risks and personal preventive measures. It is also IT's task to keep the company workers' knowledge base up-to-date on any new dangers that they should be cautious about. If the employees are educated, there will be a much lower chance that anyone will accidentally cause a breach in security by not locking down their laptop or bring in a wide open home access point to extend their mobile range. Employees need to be made aware that company laptop security extends to outside of their site walls as well. This includes places such as coffee houses where workers can be at their most vulnerable. The last item on the list deals with 24/7 active defense measures to ensure that the company network is secure and compliant. This can take the form of regularly looking at access point, server, and firewall logs to try to detect any unusual activity. For instance, if any large files went through an access point in the early hours of the morning, a serious investigation into the incident would be called for. There are a number of software and hardware devices that can be used to supplement the usual logs and usual other safety measures.

RF Shielding

It's practical in some cases to apply specialized wall paint and window film to a room or building to significantly attenuate wireless signals, which keeps the signals from propagating outside a facility. This can significantly improve wireless security because it's difficult for hackers to receive the signals beyond the controlled area of an enterprise, such as within parking lots.

Denial of Service Defense

Most DoS attacks are easy to detect. However, a lot of them are difficult to stop even after detection. Here are three of the most common ways to stop a DoS attack.

- Black Holing
- Validating the Handshake
- Rate Limiting

Black Holing is one possible way of stopping a DoS attack. This is a situation where we drop all IP packets from an attacker. This is not a very good long-term strategy because attackers can change their source address very quickly.

This may have negative effects if done automatically. An attacker could knowingly spoof attack packets with the IP address of a corporate partner. Automated defenses could block legitimate traffic from that partner and cause additional problems.

Validating the Handshake involves creating false opens, and not setting aside resources until the sender acknowledges. Some firewalls address SYN floods by pre-validating the TCP handshake. This is done by creating false opens. Whenever a SYN segment arrives, the firewall sends back a SYN/ACK segment, without passing the SYN segment on to the target server.

Only when the firewall gets back an ACK, which would happen only in a legitimate connection, would the firewall send the original SYN segment on to the server for which it was originally intended. The firewall doesn't set aside resources for a connection when a SYN segment arrives, so handling a large number of false SYN segments is only a small burden.

Rate limiting can be used to reduce a certain type of traffic down to an amount the can be reasonably dealt with. Broadcasting to the internal network could still be used, but only at a limited rate for example. This is for more subtle DoS attacks. This is good if an attack is aimed at a single server because it keeps transmission lines at least partially open for other communication.

Rate limiting frustrates both the attacker, and the legitimate users. This helps but does not fully solve the problem. Once DoS traffic clogs the access line going to the internet, there is nothing a border firewall can do to help the situation. Most DoS attacks are problems of the community which can only be stopped with the help of ISP's and organizations whose computers are taken over as bots and used to attack other firms.

Mobile Devices

With increasing number of mobile devices with 802.1x interfaces, security of such mobile devices becomes a concern. While open standards such as Kismet are targeted towards securing laptops, access points solutions should extend towards covering mobile devices also. Host based solutions for mobile handsets and PDA's with 802.1x interface.

Security within mobile devices fall under three categories:

1. Protecting against ad hoc networks
2. Connecting to rogue access points
3. Mutual authentication schemes such as WPA2 as described above

Wireless IPS solutions now offer wireless security for mobile devices.

Mobile patient monitoring devices are becoming an integral part of healthcare industry and these devices will eventually become the method of choice for accessing and implementing health checks for patients located in remote areas. For these types of patient monitoring systems, security and reliability are critical, because they can influence the condition of patients, and could leave medical professionals in the dark about the condition of the patient if compromised.

Implementing Network Encryption

In order to implement 802.11i, one must first make sure both that the router/access point(s), as well as all client devices are indeed equipped to support the network encryption. If this is done, a server such as RADIUS, ADS, NDS, or LDAP needs to be integrated. This server can be a computer on the local network, an access point / router with integrated authentication server, or a remote server. AP's/routers with integrated authentication servers are often very expensive and specifically an option for commercial usage like hot spots. Hosted 802.1X servers via the Internet require a monthly fee; running a private server is free yet has the disadvantage that one must set it up and that the server needs to be on continuously.

To set up a server, server and client software must be installed. Server software required is an enterprise authentication server such as RADIUS, ADS, NDS, or LDAP. The required software can be picked from various suppliers as Microsoft, Cisco, Funk Software, Meetinghouse Data, and from some open-source projects. Software includes:

- Aradial RADIUS Server
- Cisco Secure Access Control Software
- freeRADIUS (open-source)
- Funk Software Steel Belted RADIUS (Odyssey)
- Microsoft Internet Authentication Service
- Meetinghouse Data EAGIS
- SkyFriendz (free cloud solution based on freeRADIUS)

Client software comes built-in with Windows XP and may be integrated into other OS's using any of following software:

- AEGIS-client
- Cisco ACU-client

- Intel PROSet/Wireless Software
- Odyssey client
- Xsupplicant (open1X)-project

RADIUS

Remote Authentication Dial In User Service (RADIUS) is an AAA (authentication, authorization and accounting) protocol used for remote network access. RADIUS was originally proprietary but was later published under ISOC documents RFC 2138 and RFC 2139. The idea is to have an inside server act as a gatekeeper by verifying identities through a username and password that is already pre-determined by the user. A RADIUS server can also be configured to enforce user policies and restrictions as well as record accounting information such as connection time for purposes such as billing.

Open Access Points

Today, there is almost full wireless network coverage in many urban areas - the infrastructure for the wireless community network (which some consider to be the future of the internet) is already in place. One could roam around and always be connected to Internet if the nodes were open to the public, but due to security concerns, most nodes are encrypted and the users don't know how to disable encryption. Many people consider it proper etiquette to leave access points open to the public, allowing free access to Internet. Others think the default encryption provides substantial protection at small inconvenience, against dangers of open access that they fear may be substantial even on a home DSL router.

The density of access points can even be a problem - there are a limited number of channels available, and they partly overlap. Each channel can handle multiple networks, but places with many private wireless networks (for example, apartment complexes), the limited number of Wi-Fi radio channels might cause slowness and other problems.

According to the advocates of Open Access Points, it shouldn't involve any significant risks to open up wireless networks for the public:

- The wireless network is after all confined to a small geographical area. A computer connected to the Internet and having improper configurations or other security problems can be exploited by anyone from anywhere in the world, while only clients in a small geographical range can exploit an open wireless access point. Thus the exposure is low with an open wireless access point, and the risks with having an open wireless network are small. However, one should be aware that an open wireless router will give access to the local network, often including access to file shares and printers.
- The only way to keep communication truly secure is to use end-to-end encryption. For example, when accessing an internet bank, one would almost always use strong encryption from the web browser and all the way to the bank - thus it shouldn't be risky to do banking over an unencrypted wireless network. The argument is that anyone can sniff the traffic applies to wired networks too, where system administrators and possible hackers have ac-

cess to the links and can read the traffic. Also, anyone knowing the keys for an encrypted wireless network can gain access to the data being transferred over the network.

- If services like file shares, access to printers etc. are available on the local net, it is advisable to have authentication (i.e. by password) for accessing it (one should never assume that the private network is not accessible from the outside). Correctly set up, it should be safe to allow access to the local network to outsiders.
- With the most popular encryption algorithms today, a sniffer will usually be able to compute the network key in a few minutes.
- It is very common to pay a fixed monthly fee for the Internet connection, and not for the traffic - thus extra traffic will not be detrimental.
- Where Internet connections are plentiful and cheap, freeloaders will seldom be a prominent nuisance.

On the other hand, in some countries including Germany, persons providing an open access point may be made (partially) liable for any illegal activity conducted via this access point. Also, many contracts with ISPs specify that the connection may not be shared with other persons.

Wireless Intrusion Prevention System

In computing, a wireless intrusion prevention system (WIPS) is a network device that monitors the radio spectrum for the presence of unauthorized access points (intrusion detection), and can automatically take countermeasures (intrusion prevention).

Purpose

The primary purpose of a WIPS is to prevent unauthorized network access to local area networks and other information assets by wireless devices. These systems are typically implemented as an overlay to an existing Wireless LAN infrastructure, although they may be deployed standalone to enforce no-wireless policies within an organization. Some advanced wireless infrastructure has integrated WIPS capabilities.

Large organizations with many employees are particularly vulnerable to security breaches caused by rogue access points. If an employee (trusted entity) in a location brings in an easily available wireless router, the entire network can be exposed to anyone within range of the signals.

In July 2009, the PCI Security Standards Council published wireless guidelines for PCI DSS recommending the use of WIPS to automate wireless scanning for large organizations.

Intrusion Detection

A wireless intrusion detection system (WIDS) monitors the radio spectrum for the presence of unauthorized, rogue access points and the use of wireless attack tools. The system monitors the radio spectrum used by wireless LANs, and immediately alerts a systems administrator whenever a rogue access point is detected. Conventionally it is achieved by comparing the MAC address of the participating wireless devices.

Rogue devices can spoof MAC address of an authorized network device as their own. New research uses fingerprinting approach to weed out devices with spoofed MAC addresses. The idea is to compare the unique signatures exhibited by the signals emitted by each wireless device against the known signatures of pre-authorized, known wireless devices.

Intrusion Prevention

In addition to intrusion detection, a WIPS also includes features that prevent against the threat *automatically*. For automatic prevention, it is required that the WIPS is able to accurately detect and automatically classify a threat.

The following types of threats can be prevented by a good WIPS:

- Rogue access points – WIPS should understand the difference between rogue APs and external (neighbor's) APs
- Mis-configured AP
- Client mis-association
- Unauthorized association
- Man-in-the-middle attack
- *Ad hoc* networks
- MAC spoofing
- Honeypot / evil twin attack
- Denial-of-service attack

Implementation

WIPS configurations consist of three components:

- Sensors — These devices contain antennas and radios that scan the wireless spectrum for packets and are installed throughout areas to be protected
- Server — The WIPS server centrally analyzes packets captured by sensors
- Console — The console provides the primary user interface into the system for administration and reporting

A simple intrusion detection system can be a single computer, connected to a wireless signal processing device, and antennas placed throughout the facility. For huge organizations, a Multi Network Controller provides central control of multiple WIPS servers, while for SOHO or SMB customers, all the functionality of WIPS is available in single box.

In a WIPS implementation, users first define the operating wireless policies in the WIPS. The

WIPS sensors then analyze the traffic in the air and send this information to WIPS server. The WIPS server correlates the information, validates it against the defined policies, and classifies if it is a threat. The administrator of the WIPS is then notified of the threat, or, if a policy has been set accordingly, the WIPS takes automatic protection measures.

WIPS is configured as either a network implementation or a hosted implementation.

Network Implementation

In a network WIPS implementation, server, sensors and the console are all placed inside a private network and are not accessible from the Internet.

Sensors communicate with the server over a private network using a private port. Since the server resides on the private network, users can access the console only from within the private network.

A network implementation is suitable for organizations where all locations are within the private network.

Hosted Implementation

In a hosted WIPS implementation, sensors are installed inside a private network. However, the server is hosted in secure data center and is accessible on the Internet. Users can access the WIPS console from anywhere on the Internet. A hosted WIPS implementation is as secure as a network implementation because the data flow is encrypted between sensors and server, as well as between server and console. A hosted WIPS implementation requires very little configuration because the sensors are programmed to automatically look for the server on the Internet over a secure TLS connection.

For a large organization with locations that are not a part of a private network, a hosted WIPS implementation simplifies deployment significantly because sensors connect to the Server over the Internet without requiring any special configuration. Additionally, the Console can be accessed securely from anywhere on the Internet.

Hosted WIPS implementations are available in an on-demand, subscription-based software as a service model. Hosted implementations may be appropriate for organizations looking to fulfill the minimum scanning requirements of PCI DSS.

Intrusion Detection System

An intrusion detection system (IDS) is a device or software application that monitors a network or systems for malicious activity or policy violations. Any detected activity or violation is typically reported either to an administrator or collected centrally using a security information and event management (SIEM) system. A SIEM system combines outputs from multiple sources, and uses alarm filtering techniques to distinguish malicious activity from false alarms.

There is a wide spectrum of IDS, varying from antivirus software to hierarchical systems that monitor the traffic of an entire backbone network. The most common classifications are network intrusion detection systems (NIDS) and host-based intrusion detection systems (HIDS). A system that monitors important operating system files is an example of a HIDS, while a system that analyzes incoming network traffic is an example of a NIDS. It is also possible to classify IDS by detection approach: the most well-known variants are signature-based detection (recognizing bad patterns, such as malware) and anomaly-based detection (detecting deviations from a model of "good" traffic, which often relies on machine learning). Some IDS have the ability to respond to detected intrusions. Systems with response capabilities are typically referred to as an intrusion prevention system.

Comparison with Firewalls

Though they both relate to network security, an intrusion detection system (IDS) differs from a firewall in that a firewall looks outwardly for intrusions in order to stop them from happening. Firewalls limit access between networks to prevent intrusion and do not signal an attack from inside the network. An IDS evaluates a suspected intrusion once it has taken place and signals an alarm. An IDS also watches for attacks that originate from within a system. This is traditionally achieved by examining network communications, identifying heuristics and patterns (often known as signatures) of common computer attacks, and taking action to alert operators. A system that terminates connections is called an intrusion prevention system, and is another form of an application layer firewall.

Classifications

IDS can be classified by where detection takes place (network or host) and the detection method that is employed.

Analyzed Activity

Network Intrusion Detection Systems

Network intrusion detection systems (NIDS) are placed at a strategic point or points within the network to monitor traffic to and from all devices on the network. It performs an analysis of passing traffic on the entire subnet, and matches the traffic that is passed on the subnets to the library of known attacks. Once an attack is identified, or abnormal behavior is sensed, the alert can be sent to the administrator. An example of an NIDS would be installing it on the subnet where firewalls are located in order to see if someone is trying to break into the firewall. Ideally one would scan all inbound and outbound traffic, however doing so might create a bottleneck that would impair the overall speed of the network. OPNET and NetSim are commonly used tools for simulation network intrusion detection systems. NID Systems are also capable of comparing signatures for similar packets to link and drop harmful detected packets which have a signature matching the records in the NIDS. When we classify the designing of the NIDS according to the system interactivity property, there are two types: on-line and off-line NIDS. On-line NIDS deals with the network in real time. It analyses the Ethernet packets and applies some rules, to decide if it is an attack or not. Off-line NIDS deals with stored data and passes it through some processes to decide if it is an attack or not.

Host Intrusion Detection Systems

Host intrusion detection systems (HIDS) run on individual hosts or devices on the network. A

HIDS monitors the inbound and outbound packets from the device only and will alert the user or administrator if suspicious activity is detected. It takes a snapshot of existing system files and matches it to the previous snapshot. If the critical system files were modified or deleted, an alert is sent to the administrator to investigate. An example of HIDS usage can be seen on mission critical machines, which are not expected to change their configurations.

Intrusion detection systems can also be system-specific using custom tools and honeypots.

Detection Method

Signature-based

Signature-based IDS refers to the detection of attacks by looking for specific patterns, such as byte sequences in network traffic, or known malicious instruction sequences used by malware. This terminology originates from anti-virus software, which refers to these detected patterns as signatures. Although signature-based IDS can easily detect known attacks, it is impossible to detect new attacks, for which no pattern is available.

Anomaly-based

Anomaly-based intrusion detection systems were primarily introduced to detect unknown attacks, in part due to the rapid development of malware. The basic approach is to use machine learning to create a model of trustworthy activity, and then compare new behavior against this model. Although this approach enables the detection of previously unknown attacks, it suffers from false positives: previously unknown legitimate activity may also be classified as malicious.

Intrusion Prevention

Some systems may attempt to stop an intrusion attempt but this is neither required nor expected of a monitoring system. Intrusion detection and prevention systems (IDPS) are primarily focused on identifying possible incidents, logging information about them, and reporting attempts. In addition, organizations use IDPSes for other purposes, such as identifying problems with security policies, documenting existing threats and deterring individuals from violating security policies. IDPSes have become a necessary addition to the security infrastructure of nearly every organization.

IDPSes typically record information related to observed events, notify security administrators of important observed events and produce reports. Many IDPSes can also respond to a detected threat by attempting to prevent it from succeeding. They use several response techniques, which involve the IDPS stopping the attack itself, changing the security environment (e.g. reconfiguring a firewall) or changing the attack's content.

Intrusion prevention systems (IPS), also known as intrusion detection and prevention systems (IDPS), are network security appliances that monitor network or system activities for malicious activity. The main functions of intrusion prevention systems are to identify malicious activity, log information about this activity, report it and attempt to block or stop it..

Intrusion prevention systems are considered extensions of intrusion detection systems because they both monitor network traffic and/or system activities for malicious activity. The main differences are, unlike intrusion detection systems, intrusion prevention systems are placed in-line and are able to actively prevent or block intrusions that are detected. IPS can take such actions as sending an alarm, dropping detected malicious packets, resetting a connection or blocking traffic from the offending IP address. An IPS also can correct cyclic redundancy check (CRC) errors, defragment packet streams, mitigate TCP sequencing issues, and clean up unwanted transport and network layer options.

Classification

Intrusion prevention systems can be classified into four different types:

1. Network-based intrusion prevention system (NIPS): monitors the entire network for suspicious traffic by analyzing protocol activity.
2. Wireless intrusion prevention systems (WIPS): monitor a wireless network for suspicious traffic by analyzing wireless networking protocols.
3. Network behavior analysis (NBA): examines network traffic to identify threats that generate unusual traffic flows, such as distributed denial of service (DDoS) attacks, certain forms of malware and policy violations.
4. Host-based intrusion prevention system (HIPS): an installed software package which monitors a single host for suspicious activity by analyzing events occurring within that host.

Detection Methods

The majority of intrusion prevention systems utilize one of three detection methods: signature-based, statistical anomaly-based, and stateful protocol analysis.

1. Signature-Based Detection: Signature based IDS monitors packets in the Network and compares with pre-configured and pre-determined attack patterns known as signatures.
2. Statistical anomaly-based detection: An IDS which is anomaly based will monitor network traffic and compare it against an established baseline. The baseline will identify what is "normal" for that network – what sort of bandwidth is generally used, what protocols are used that it may raise a False Positive alarm for a legitimate use of bandwidth if the baselines are not intelligently configured.
3. Stateful Protocol Analysis Detection: This method identifies deviations of protocol states by comparing observed events with "predetermined profiles of generally accepted definitions of benign activity."

Limitations

- Noise can severely limit an intrusion detection system's effectiveness. Bad packets generated from software bugs, corrupt DNS data, and local packets that escaped can create a significantly high false-alarm rate.
- It is not uncommon for the number of real attacks to be far below the number of false-

alarms. Number of real attacks is often so far below the number of false-alarms that the real attacks are often missed and ignored.

- Many attacks are geared for specific versions of software that are usually outdated. A constantly changing library of signatures is needed to mitigate threats. Outdated signature databases can leave the IDS vulnerable to newer strategies.
- For signature-based IDSes there will be lag between a new threat discovery and its signature being applied to the IDS. During this lag time the IDS will be unable to identify the threat.
- It cannot compensate for a weak identification and authentication mechanisms or for weaknesses in network protocols. When an attacker gains access due to weak authentication mechanism then IDS cannot prevent the adversary from any malpractice.
- Encrypted packets are not processed by the intrusion detection software. Therefore, the encrypted packet can allow an intrusion to the network that is undiscovered until more significant network intrusions have occurred.
- Intrusion detection software provides information based on the network address that is associated with the IP packet that is sent into the network. This is beneficial if the network address contained in the IP packet is accurate. However, the address that is contained in the IP packet could be faked or scrambled.
- Due to the nature of NIDS systems, and the need for them to analyse protocols as they are captured, NIDS systems can be susceptible to same protocol based attacks that network hosts may be vulnerable. Invalid data and TCP/IP stack attacks may cause an NIDS to crash.

Evasion Techniques

There are a number of techniques which attackers are using, the following are considered 'simple' measures which can be taken to evade IDS:

- Fragmentation: by sending fragmented packets, the attacker will be under the radar and can easily bypass the detection system's ability to detect the attack signature.
- Avoiding defaults: The TCP port utilised by a protocol does not always provide an indication to the protocol which is being transported. For example, an IDS may expect to detect a trojan on port 12345. If an attacker had reconfigured it to use a different port the IDS may not be able to detect the presence of the trojan.
- Coordinated, low-bandwidth attacks: coordinating a scan among numerous attackers (or agents) and allocating different ports or hosts to different attackers makes it difficult for the IDS to correlate the captured packets and deduce that a network scan is in progress.
- Address spoofing/proxying: attackers can increase the difficulty of the ability of Security Administrators to determine the source of the attack by using poorly secured or incorrectly configured proxy servers to bounce an attack. If the source is spoofed and bounced by a server then it makes it very difficult for IDS to detect the origin of the attack.

- Pattern change evasion: IDSs generally rely on 'pattern matching' to detect an attack. By changing the data used in the attack slightly, it may be possible to evade detection. For example, an IMAP server may be vulnerable to a buffer overflow, and an IDS is able to detect the attack signature of 10 common attack tools. By modifying the payload sent by the tool, so that it does not resemble the data that the IDS expects, it may be possible to evade detection.

Development

One preliminary IDS concept consisted of a set of tools intended to help administrators review audit trails. User access logs, file access logs, and system event logs are examples of audit trails.

Fred Cohen noted in 1987 that it is impossible to detect an intrusion in every case, and that the resources needed to detect intrusions grow with the amount of usage.

Dorothy E. Denning, assisted by Peter G. Neumann, published a model of an IDS in 1986 that formed the basis for many systems today. Her model used statistics for anomaly detection, and resulted in an early IDS at SRI International named the Intrusion Detection Expert System (IDES), which ran on Sun workstations and could consider both user and network level data. IDES had a dual approach with a rule-based Expert System to detect known types of intrusions plus a statistical anomaly detection component based on profiles of users, host systems, and target systems. Lunt proposed adding an Artificial neural network as a third component. She said all three components could then report to a resolver. SRI followed IDES in 1993 with the Next-generation Intrusion Detection Expert System (NIDES).

The Multics intrusion detection and alerting system (MIDAS), an expert system using P-BEST and Lisp, was developed in 1988 based on the work of Denning and Neumann. Haystack was also developed in that year using statistics to reduce audit trails.

Wisdom & Sense (W&S) was a statistics-based anomaly detector developed in 1989 at the Los Alamos National Laboratory. W&S created rules based on statistical analysis, and then used those rules for anomaly detection.

In 1990, the Time-based Inductive Machine (TIM) did anomaly detection using inductive learning of sequential user patterns in Common Lisp on a VAX 3500 computer. The Network Security Monitor (NSM) performed masking on access matrices for anomaly detection on a Sun-3/50 workstation. The Information Security Officer's Assistant (ISOA) was a 1990 prototype that considered a variety of strategies including statistics, a profile checker, and an expert system. ComputerWatch at AT&T Bell Labs used statistics and rules for audit data reduction and intrusion detection.

Then, in 1991, researchers at the University of California, Davis created a prototype Distributed Intrusion Detection System (DIDS), which was also an expert system. The Network Anomaly Detection and Intrusion Reporter (NADIR), also in 1991, was a prototype IDS developed at the Los Alamos National Laboratory's Integrated Computing Network (ICN), and was heavily influenced by the work of Denning and Lunt. NADIR used a statistics-based anomaly detector and an expert system.

The Lawrence Berkeley National Laboratory announced Bro in 1998, which used its own rule language for packet analysis from libpcap data. Network Flight Recorder (NFR) in 1999 also used libpcap. APE was developed as a packet sniffer, also using libpcap, in November, 1998, and was renamed Snort one month later. APE has since become the world's largest used IDS/IPS system with over 300,000 active users.

The Audit Data Analysis and Mining (ADAM) IDS in 2001 used tcpdump to build profiles of rules for classifications.

In 2003, Yongguang Zhang and Wenke Lee argue for the importance of IDS in networks with mobile nodes.

RADIUS

Remote Authentication Dial-In User Service (RADIUS) is a networking protocol that provides centralized Authentication, Authorization, and Accounting (AAA or Triple A) management for users who connect and use a network service. RADIUS was developed by Livingston Enterprises, Inc. in 1991 as an access server authentication and accounting protocol and later brought into the Internet Engineering Task Force (IETF) standards.

Because of the broad support and the ubiquitous nature of the RADIUS protocol, it is often used by ISPs and enterprises to manage access to the Internet or internal networks, wireless networks, and integrated e-mail services. These networks may incorporate modems, DSL, access points, VPNs, network ports, web servers, etc.

RADIUS is a client/server protocol that runs in the application layer, and can use either TCP or UDP as transport. Network access servers, the gateways that control access to a network, usually contain a RADIUS client component that communicates with the RADIUS server. RADIUS is often the back-end of choice for 802.1X authentication as well.

The RADIUS server is usually a background process running on a UNIX or Microsoft Windows server.

Protocol Components

RADIUS is an AAA protocol which manages network access in the following two-step process, also known as a *AAA transaction*. AAA stands for authentication, authorization and accounting. Authentication and authorization characteristics in RADIUS are described in RFC 2865 while accounting is described by RFC 2866.

Authentication and Authorization

The user or machine sends a request to a Network Access Server (NAS) to gain access to a particular network resource using access credentials. The credentials are passed to the NAS device via the link-layer protocol - for example, Point-to-Point Protocol (PPP) in the case of many dialup or DSL providers or posted in an HTTPS secure web form.

In turn, the NAS sends a RADIUS *Access Request* message to the RADIUS server, requesting authorization to grant access via the RADIUS protocol.

This request includes access credentials, typically in the form of username and password or security certificate provided by the user. Additionally, the request may contain other information which the NAS knows about the user, such as its network address or phone number, and information regarding the user's physical point of attachment to the NAS.

The RADIUS server checks that the information is correct using authentication schemes such as PAP, CHAP or EAP. The user's proof of identification is verified, along with, optionally, other information related to the request, such as the user's network address or phone number, account status, and specific network service access privileges. Historically, RADIUS servers checked the user's information against a locally stored flat file database. Modern RADIUS servers can do this, or can refer to external sources — commonly SQL, Kerberos, LDAP, or Active Directory servers — to verify the user's credentials.

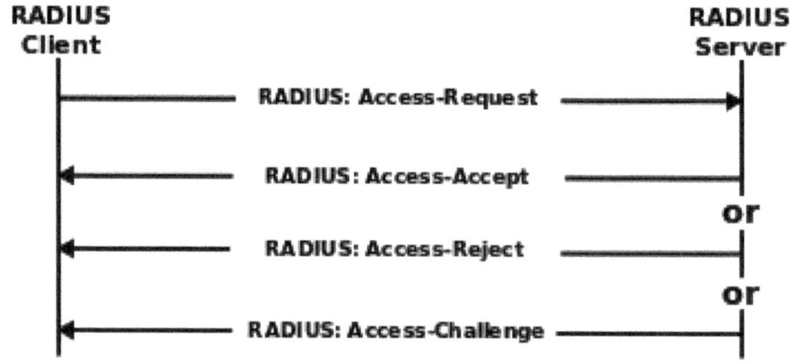

RADIUS Authentication and Authorization Flow

The RADIUS server then returns one of three responses to the NAS : 1) Access Reject, 2) Access Challenge, or 3) Access Accept.

- *Access Reject* - The user is unconditionally denied access to all requested network resources. Reasons may include failure to provide proof of identification or an unknown or inactive user account.
- *Access Challenge* - Requests additional information from the user such as a secondary password, PIN, token, or card. Access Challenge is also used in more complex authentication dialogs where a secure tunnel is established between the user machine and the Radius Server in a way that the access credentials are hidden from the NAS.
- *Access Accept* - The user is granted access. Once the user is authenticated, the RADIUS server will often check that the user is authorized to use the network service requested. A given user may be allowed to use a company's wireless network, but not its VPN service, for example. Again, this information may be stored locally on the RADIUS server, or may be looked up in an external source such as LDAP or Active Directory.

Each of these three RADIUS responses may include a Reply-Message attribute which may give a reason for the rejection, the prompt for the challenge, or a welcome message for the accept. The text in the attribute can be passed on to the user in a return web page.

Authorization attributes are conveyed to the NAS stipulating terms of access to be granted. For example, the following authorization attributes may be included in an Access-Accept:

- The specific IP address to be assigned to the user
- The address pool from which the user's IP should be chosen
- The maximum length of time that the user may remain connected
- An access list, priority queue or other restrictions on a user's access
- L2TP parameters
- VLAN parameters
- Quality of Service (QoS) parameters

When a client is configured to use RADIUS, any user of the client presents authentication information to the client. This might be with a customizable login prompt, where the user is expected to enter their username and password. Alternatively, the user might use a link framing protocol such as the Point-to-Point Protocol (PPP), which has authentication packets which carry this information.

Once the client has obtained such information, it may choose to authenticate using RADIUS. To do so, the client creates an "Access- Request" containing such Attributes as the user's name, the user's password, the ID of the client and the Port ID which the user is accessing. When a password is present, it is hidden using a method based on the RSA Message Digest Algorithm MD5.

Accounting

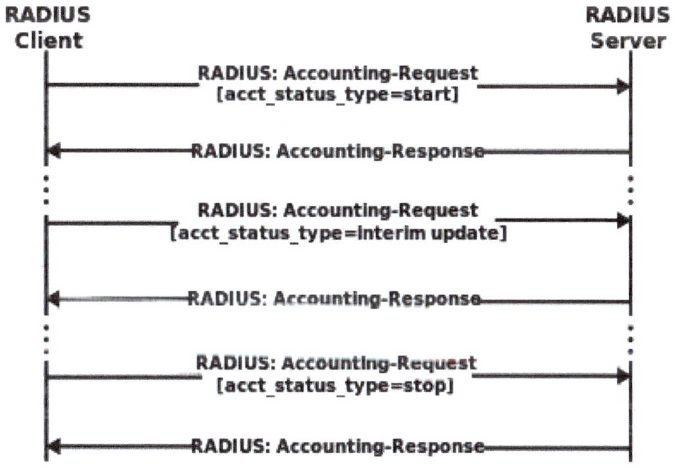

RADIUS Accounting Flow

Accounting is described in RFC 2866.

When network access is granted to the user by the NAS, an *Accounting Start* (a RADIUS Accounting Request packet containing an Acct-Status-Type attribute with the value "start") is sent by the NAS to the RADIUS server to signal the start of the user's network access. "Start" records typically contain the user's identification, network address, point of attachment and a unique session identifier.

Periodically, *Interim Update* records (a RADIUS Accounting Request packet containing an Acct-Status-Type attribute with the value "interim-update") may be sent by the NAS to the RADIUS server, to update it on the status of an active session. "Interim" records typically convey the current session duration and information on current data usage.

Finally, when the user's network access is closed, the NAS issues a final *Accounting Stop* record (a RADIUS Accounting Request packet containing an Acct-Status-Type attribute with the value "stop") to the RADIUS server, providing information on the final usage in terms of time, packets transferred, data transferred, reason for disconnect and other information related to the user's network access.

Typically, the client sends Accounting-Request packets until it receives an Accounting-Response acknowledgement, using some retry interval.

The primary purpose of this data is that the user can be billed accordingly; the data is also commonly used for statistical purposes and for general network monitoring.

Roaming

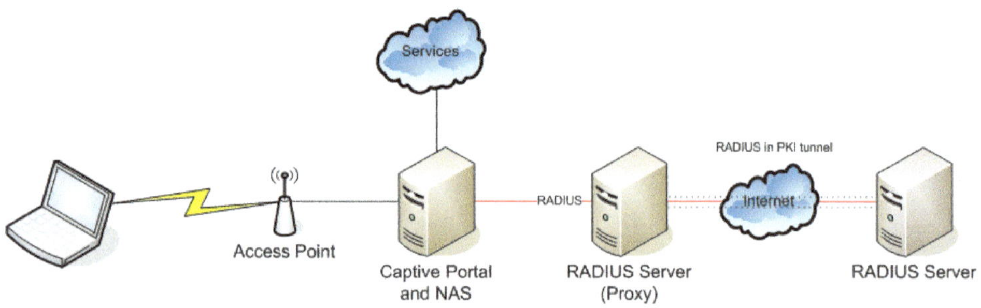

Roaming using a proxy RADIUS AAA server.

RADIUS is commonly used to facilitate roaming between ISPs, for example:

- by companies which provide a single global set of credentials that are usable on many public networks;
- by independent, but collaborating, institutions issuing their own credentials to their own users, that allow a visitor from one to another to be authenticated by their home institution, such as in eduroam.

RADIUS facilitates this by the use of *realms*, which identify where the RADIUS server should forward the AAA requests for processing.

Realms

A realm is commonly appended to a user's user name and delimited with an '@' sign, resembling an email address domain name. This is known as *postfix* notation for the realm. Another common usage is *prefix* notation, which involves prepending the realm to the username and using '\' as a delimiter. Modern RADIUS servers allow any character to be used as a realm delimiter, although in practice '@' and '\' are usually used.

Realms can also be compounded using both prefix and postfix notation, to allow for complicated roaming scenarios; for example, somedomain.com\username@anotherdomain.com could be a valid username with two realms.

Although realms often resemble domains, it is important to note that realms are in fact arbitrary text and need not contain real domain names. Realm formats are standardized in RFC 4282, which defines a Network Access Identifier (NAI) in the form of 'user@realm'. In that specification, the 'realm' portion is required to be a domain name. However, this practice is not always followed. RFC 7542 replaced RFC 4282 in May 2015.

Proxy Operations

When a RADIUS server receives an AAA request for a user name containing a realm, the server will reference a table of configured realms. If the realm is known, the server will then *proxy* the request to the configured home server for that domain. The behavior of the proxying server regarding the removal of the realm from the request ("stripping") is configuration-dependent on most servers. In addition, the proxying server can be configured to add, remove or rewrite AAA requests when they are proxied over time again.

Proxy Chaining is possible in RADIUS and authentication/authorization and accounting packets are usually routed between a NAS Device and a Home server through a series of proxies. Some of advantages of using Proxy chains include scalability improvements, policy implementations and capability adjustments. But in roaming scenarios, the NAS, Proxies and Home Server could be typically managed by different administrative entities. Hence, the trust factor among the proxies gains more significance under such Inter-domain applications. Further, the absence of end to end security in RADIUS adds to the criticality of trust among the Proxies involved. Proxy Chains are explained in RFC 2607.

Security

Roaming with RADIUS exposes the users to various security and privacy concerns. More generally, some roaming partners establish a secure tunnel between the RADIUS servers to ensure that users' credentials cannot be intercepted while being proxied across the internet. This is a concern as the MD5 hash built into RADIUS is considered insecure.

Packet Structure

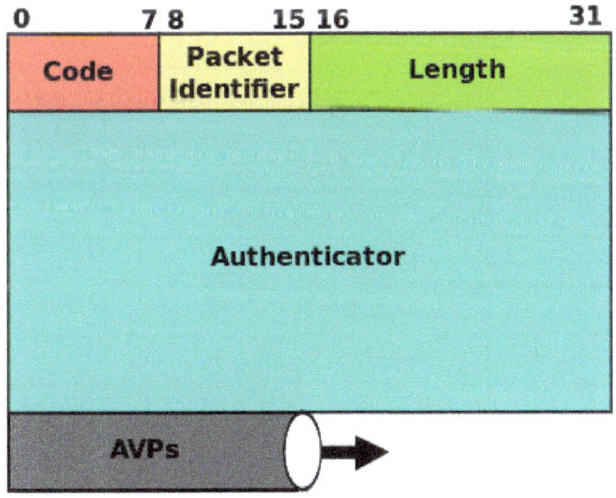

RADIUS packet data format.

The RADIUS packet data format is shown to the right. The fields are transmitted from left to right, starting with the code, the identifier, the length, the authenticator and the attributes.

RADIUS Codes (decimal) are assigned as follows:

Code	Assignment
1	Access-Request
2	Access-Accept
3	Access-Reject
4	Accounting-Request
5	Accounting-Response
11	Access-Challenge
12	Status-Server (experimental)
13	Status-Client (experimental)
255	Reserved

The Identifier field aids in matching requests and replies.

The Length field indicates the length of the entire RADIUS packet including the Code, Identifier, Length, Authenticator and optional Attribute fields.

The Authenticator is used to authenticate the reply from the RADIUS server, and is used in encrypting passwords; its length is 16 bytes.

Attribute Value Pairs

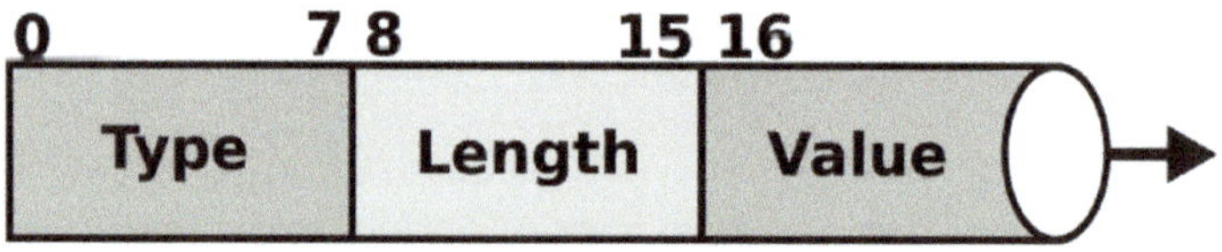

RADIUS AVP layout

The RADIUS Attribute Value Pairs (AVP) carry data in both the request and the response for the authentication, authorization, and accounting transactions. The length of the radius packet is used to determine the end of the AVPs.

AVP Type	Assignment
1	User-Name
2	User-Password
3	CHAP-Password
4	NAS-IP-Address
5	NAS-Port
6	Service-Type
7	Framed-Protocol
8	Framed-IP-Address
9	Framed-IP-Netmask

10	Framed-Routing
11	Filter-Id
12	Framed-MTU
13	Framed-Compression
14	Login-IP-Host
15	Login-Service
16	Login-TCP-Port
18	Reply-Message
19	Callback-Number
20	Callback-Id
22	Framed-Route
23	Framed-IPX-Network
24	State
25	Class
26	Vendor-Specific
27	Session-Timeout
28	Idle-Timeout
29	Termination-Action
30	Called-Station-Id
31	Calling-Station-Id
32	NAS-Identifier
33	Proxy-State
34	Login-LAT-Service
35	Login-LAT-Node
36	Login-LAT-Group
37	Framed-AppleTalk-Link
38	Framed-AppleTalk-Network
39	Framed-AppleTalk-Zone
40	Acct-Status-Type
41	Acct-Delay-Time
42	Acct-Input-Octets
43	Acct-Output-Octets
44	Acct-Session-Id
45	Acct-Authentic
46	Acct-Session-Time
47	Acct-Input-Packets
48	Acct-Output-Packets
49	Acct-Terminate-Cause
50	Acct-Multi-Session-Id
51	Acct-Link-Count

52	Acct-Input-Gigawords
53	Acct-Output-Gigawords
55	Event-Timestamp
56	Egress-VLANID
57	Ingress-Filters
58	Egress-VLAN-Name
59	User-Priority-Table
60	CHAP-Challenge
61	NAS-Port-Type
62	Port-Limit
63	Login-LAT-Port
64	Tunnel-Type
65	Tunnel-Medium-Type
66	Tunnel-Client-Endpoint
67	Tunnel-Server-Endpoint
68	Acct-Tunnel-Connection
69	Tunnel-Password
70	ARAP-Password
71	ARAP-Features
72	ARAP-Zone-Access
73	ARAP-Security
74	ARAP-Security-Data
75	Password-Retry
76	Prompt
77	Connect-Info
78	Configuration-Token
79	EAP-Message
80	Message-Authenticator
81	Tunnel-Private-Group-ID
82	Tunnel-Assignment-ID
83	Tunnel-Preference
84	ARAP-Challenge-Response
85	Acct-Interim-Interval
86	Acct-Tunnel-Packets-Lost
87	NAS-Port-Id
88	Framed-Pool
89	CUI
90	Tunnel-Client-Auth-ID
91	Tunnel-Server-Auth-ID
92	NAS-Filter-Rule

94	Originating-Line-Info
95	NAS-IPv6-Address
96	Framed-Interface-Id
97	Framed-IPv6-Prefix
98	Login-IPv6-Host
99	Framed-IPv6-Route
100	Framed-IPv6-Pool
101	Error-Cause Attribute
102	EAP-Key-Name
103	Digest-Response
104	Digest-Realm
105	Digest-Nonce
106	Digest-Response-Auth
107	Digest-Nextnonce
108	Digest-Method
109	Digest-URI
110	Digest-Qop
111	Digest-Algorithm
112	Digest-Entity-Body-Hash
113	Digest-CNonce
114	Digest-Nonce-Count
115	Digest-Username
116	Digest-Opaque
117	Digest-Auth-Param
118	Digest-AKA-Auts
119	Digest-Domain
120	Digest-Stale
121	Digest-HA1
122	SIP-AOR
123	Delegated-IPv6-Prefix
124	MIP6-Feature-Vector
125	MIP6-Home-Link-Prefix
126	Operator-Name
127	Location-Information
128	Location-Data
129	Basic-Location-Policy-Rules
130	Extended-Location-Policy-Rules
131	Location-Capable
132	Requested-Location-Info
133	Framed-Management-Protocol

134	Management-Transport-Protection
135	Management-Policy-Id
136	Management-Privilege-Level
137	PKM-SS-Cert
138	PKM-CA-Cert
139	PKM-Config-Settings
140	PKM-Cryptosuite-List
141	PKM-SAID
142	PKM-SA-Descriptor
143	PKM-Auth-Key
144	DS-Lite-Tunnel-Name
145	Mobile-Node-Identifier
146	Service-Selection
147	PMIP6-Home-LMA-IPv6-Address
148	PMIP6-Visited-LMA-IPv6-Address
149	PMIP6-Home-LMA-IPv4-Address
150	PMIP6-Visited-LMA-IPv4-Address
151	PMIP6-Home-HN-Prefix
152	PMIP6-Visited-HN-Prefix
153	PMIP6-Home-Interface-ID
154	PMIP6-Visited-Interface-ID
155	PMIP6-Home-IPv4-HoA
156	PMIP6-Visited-IPv4-HoA
157	PMIP6-Home-DHCP4-Server-Address
158	PMIP6-Visited-DHCP4-Server-Address
159	PMIP6-Home-DHCP6-Server-Address
160	PMIP6-Visited-DHCP6-Server-Address
161	PMIP6-Home-IPv4-Gateway
162	PMIP6-Visited-IPv4-Gateway
163	EAP-Lower-Layer
164	GSS-Acceptor-Service-Name
165	GSS-Acceptor-Host-Name
166	GSS-Acceptor-Service-Specifics
167	GSS-Acceptor-Realm-Name
168	Framed-IPv6-Address
169	DNS-Server-IPv6-Address
170	Route-IPv6-Information
171	Delegated-IPv6-Prefix-Pool
172	Stateful-IPv6-Address-Pool
173	IPv6-6rd-Configuration

174	Allowed-Called-Station-Id
175	EAP-Peer-Id
176	EAP-Server-Id
177	Mobility-Domain-Id
178	Preauth-Timeout
179	Network-Id-Name
180	EAPoL-Announcement
181	WLAN-HESSID
182	WLAN-Venue-Info
183	WLAN-Venue-Language
184	WLAN-Venue-Name
185	WLAN-Reason-Code
186	WLAN-Pairwise-Cipher
187	WLAN-Group-Cipher
188	WLAN-AKM-Suite
189	WLAN-Group-Mgmt-Cipher
190	WLAN-RF-Band

Vendor-specific Attributes

RADIUS is extensible; many vendors of RADIUS hardware and software implement their own variants using Vendor-Specific Attributes (VSAs). Microsoft has published some of their VSAs. VSA definitions from many other companies remain proprietary and/or ad-hoc, nonetheless many VSA dictionaries can be found by downloading the source code of open source RADIUS implementations, for example FreeRADIUS or openRADIUS.

Security

The RADIUS protocol transmits obfuscated passwords using a shared secret and the MD5 hashing algorithm. As this particular implementation provides only weak protection of the user's credentials, additional protection, such as IPsec tunnels or physically secured data-center networks, should be used to further protect the RADIUS traffic between the NAS device and the RADIUS server. Additionally, the user's security credentials are the only part protected by RADIUS itself, yet other user-specific attributes such as tunnel-group IDs or vlan memberships passed over RADIUS may be considered sensitive (helpful to an attacker) or private (sufficient to identity the individual client) information as well. The RadSec protocol claims to solve aforementioned security issues.

History

RADIUS was originally specified in an RFI by Merit Network in 1991 to control dial-in access to NSFnet. Livingston Enterprises responded to the RFI with a description of a RADIUS server. Merit Network awarded the contract to Livingston Enterprises that delivered their PortMaster series of Network Access Servers and the initial RADIUS server to Merit. RADIUS was later (1997) published as RFC 2058 and RFC 2059 (current versions are RFC 2865 and RFC 2866).

Now, several commercial and open-source RADIUS servers exist. Features can vary, but most can look up the users in text files, LDAP servers, various databases, etc. Accounting records can be written to text files, various databases, forwarded to external servers, etc. SNMP is often used for remote monitoring and keep-alive checking of a RADIUS server. RADIUS proxy servers are used for centralized administration and can rewrite RADIUS packets on the fly (for security reasons, or to convert between vendor dialects).

The Diameter protocol was intended as the replacement for RADIUS. While both are Authentication, Authorization, and Accounting (AAA) protocols, the use-cases for the two protocols have since diverged. Diameter is largely used in the 3G space. RADIUS is used elsewhere. One of the largest barriers to having Diameter replace RADIUS is that switches and Access Points typically implement RADIUS, but not Diameter.

Diameter uses SCTP or TCP while RADIUS typically uses UDP as the transport layer. As of 2012, RADIUS can also use TCP as the transport layer with TLS for security.

Wi-Fi Protected Access

An example of a Wi-Fi Protected Access label found on a consumer device

Wi-Fi Protected Access (WPA) and Wi-Fi Protected Access II (WPA2) are two security protocols and security certification programs developed by the Wi-Fi Alliance to secure wireless computer networks. The Alliance defined these in response to serious weaknesses researchers had found in the previous system, Wired Equivalent Privacy (WEP).

WPA (sometimes referred to as the *draft IEEE 802.11i* standard) became available in 2003. The Wi-Fi Alliance intended it as an intermediate measure in anticipation of the availability of the more secure and complex WPA2. WPA2 became available in 2004 and is a common shorthand for the full IEEE 802.11i (or IEEE 802.11i-2004) standard.

A flaw in a feature added to Wi-Fi, called Wi-Fi Protected Setup, allows WPA and WPA2 security to be bypassed and effectively broken in many situations. The WPA and WPA2 security protocols implemented without using the Wi-Fi Protected Setup feature are unaffected by the security vulnerability.

WPA

The Wi-Fi Alliance intended WPA as an intermediate measure to take the place of WEP pending the availability of the full IEEE 802.11i standard. WPA could be implemented through firmware upgrades on wireless network interface cards designed for WEP that began shipping as far back as 1999. However, since the changes required in the wireless access points (APs) were more extensive than those needed on the network cards, most pre-2003 APs could not be upgraded to support WPA.

The WPA protocol implements much of the IEEE 802.11i standard. Specifically, the Temporal Key Integrity Protocol (TKIP) was adopted for WPA. WEP used a 64-bit or 128-bit encryption key that must be manually entered on wireless access points and devices and does not change. TKIP employs a per-packet key, meaning that it dynamically generates a new 128-bit key for each packet and thus prevents the types of attacks that compromised WEP.

WPA also includes a message integrity check, which is designed to prevent an attacker from altering and resending data packets. This replaces the cyclic redundancy check (CRC) that was used by the WEP standard. CRC's main flaw was that it did not provide a sufficiently strong data integrity guarantee for the packets it handled. Well tested message authentication codes existed to solve these problems, but they required too much computation to be used on old network cards. WPA uses a message integrity check algorithm called *TKIP* to verify the integrity of the packets. TKIP is much stronger than a CRC, but not as strong as the algorithm used in WPA2. Researchers have since discovered a flaw in WPA that relied on older weaknesses in WEP and the limitations of Michael to retrieve the keystream from short packets to use for re-injection and spoofing.

WPA2

WPA2 replaced WPA. WPA2, which requires testing and certification by the Wi-Fi Alliance, implements the mandatory elements of IEEE 802.11i. In particular, it includes mandatory support for CCMP, an AES-based encryption mode with strong security. Certification began in September, 2004; from March 13, 2006, WPA2 certification is mandatory for all new devices to bear the Wi-Fi trademark.

Hardware Support

WPA has been designed specifically to work with wireless hardware produced prior to the introduction of WPA protocol, which provides inadequate security through WEP. Some of these devices support WPA only after applying firmware upgrades, which are not available for some legacy devices.

Wi-Fi devices certified since 2006 support both the WPA and WPA2 security protocols. WPA2 may not work with some older network cards.

WPA Terminology

Different WPA versions and protection mechanisms can be distinguished based on the target end-user (according to the method of authentication key distribution), and the encryption protocol used.

Target Users (Authentication Key Distribution)

WPA-Personal

Also referred to as *WPA-PSK* (pre-shared key) mode, this is designed for home and small office networks and doesn't require an authentication server. Each wireless network device encrypts the network traffic using a 256 bit key. This key may be entered either as a string of 64 hexadecimal digits, or as a passphrase of 8 to 63 printable ASCII characters. If ASCII characters are used, the 256 bit key is calculated by applying the PBKDF2 key derivation function to the passphrase, using the SSID as the salt and 4096 iterations of HMAC-SHA1. WPA-Personal mode is available with both WPA and WPA2.

WPA-Enterprise

Also referred to as *WPA-802.1X mode*, and sometimes just *WPA* (as opposed to WPA-PSK), this is designed for enterprise networks and requires a RADIUS authentication server. This requires a more complicated setup, but provides additional security (e.g. protection against dictionary attacks on short passwords). Various kinds of the Extensible Authentication Protocol (EAP) are used for authentication. WPA-Enterprise mode is available with both WPA and WPA2.

Wi-Fi Protected Setup (WPS)

This is an alternative authentication key distribution method intended to simplify and strengthen the process, but which, as widely implemented, creates a major security hole via WPS PIN recovery.

Encryption Protocol

TKIP (Temporal Key Integrity Protocol)

The RC4 stream cipher is used with a 128-bit per-packet key, meaning that it dynamically generates a new key for each packet. Used by WPA.

CCMP (*C*TR mode with *C*BC-*MAC P*rotocol)

The protocol used by WPA2, based on the Advanced Encryption Standard (AES) cipher along with strong message authenticity and integrity checking that is significantly stronger in protection for both privacy and integrity than the RC4-based TKIP used by WPA. Among informal names are "AES" and "AES-CCMP". According to the 802.11n specification, this encryption protocol must be used to achieve the fast 802.11n high bitrate schemes, though not all implementations enforce this. Otherwise, the data rate will not exceed 54 MBit/s.

EAP Extensions Under WPA and WPA2 Enterprise

Originally, only EAP-TLS (Extensible Authentication Protocol - Transport Layer Security) was certified by the Wi-Fi alliance. In April 2010, the Wi-Fi Alliance announced the inclusion of additional EAP types to its WPA- and WPA2- Enterprise certification programs. This was to ensure that WPA-Enterprise certified products can interoperate with one another.

As of 2010 the certification program includes the following EAP types:

- EAP-TLS (previously tested)
- EAP-TTLS/MSCHAPv2 (April 2005)
- PEAPv0/EAP-MSCHAPv2 (April 2005)
- PEAPv1/EAP-GTC (April 2005)
- PEAP-TLS
- EAP-SIM (April 2005)
- EAP-AKA (April 2009)
- EAP-FAST (April 2009)

802.1X clients and servers developed by specific firms may support other EAP types. This certification is an attempt for popular EAP types to interoperate; their failure to do so as of 2013 is one of the major issues preventing rollout of 802.1X on heterogeneous networks.

Commercial 802.1X servers include Microsoft Internet Authentication Service and Juniper Networks Steelbelted RADIUS as well as Aradial Radius server. FreeRADIUS is an open source 802.1X server.

Security Issues

Weak Password

Pre-shared key WPA and WPA2 remain vulnerable to password cracking attacks if users rely on a weak password or passphrase. To protect against a brute force attack, a truly random passphrase of 20 characters (selected from the set of 95 permitted characters) is probably sufficient.

Brute forcing of simple passwords can be attempted using the Aircrack Suite starting from the four-way authentication handshake exchanged during association or periodic re-authentication.

To further protect against intrusion, the network's SSID should not match any entry in the top 1,000 SSIDs as downloadable rainbow tables have been pre-generated for them and a multitude of common passwords.

WPA Packet Spoofing and Decryption

Mathy Vanhoef and Frank Piessens significantly improved upon the WPA-TKIP attacks of Erik Tews and Martin Beck. They demonstrated how to inject an arbitrary amount of packets, with each packet containing at most 112 bytes of payload. This was demonstrated by implementing a port scanner, which can be executed against any client using WPA-TKIP. Additionally they showed how to decrypt arbitrary packets sent to a client. They mentioned this can be used to hijack a TCP connection, allowing an attacker to inject malicious JavaScript when the victim visits a website. In contrast, the Beck-Tews attack could only decrypt short packets with mostly known content, such as ARP messages, and only allowed injection of 3 to 7 packets of at most 28 bytes. The Beck-Tews attack also requires Quality of Service (as defined in 802.11e) to be enabled, while the Vanhoef-Piessens attack does not. Neither attack leads to recovery of the shared session key between the client and Access Point. The authors say using a short rekeying interval can prevent some attacks but not all, and strongly recommend switching from TKIP to AES-based CCMP.

Halvorsen and others show how to modify the Beck-Tews attack to allow injection of 3 to 7 packets having a size of at most 596 bytes. The downside is that their attack requires substantially more time to execute: approximately 18 minutes and 25 seconds. In other work Vanhoef and Piessens showed that, when WPA is used to encrypt broadcast packets, their original attack can also be executed. This is an important extension, as substantially more networks use WPA to protect broadcast packets, than to protect unicast packets. The execution time of this attack is on average around 7 minutes, compared to the 14 minutes of the original Vanhoef-Piessens and Beck-Tews attack.

The vulnerabilities of TKIP are significant in that WPA-TKIP had been held to be an extremely safe combination; indeed, WPA-TKIP is still a configuration option upon a wide variety of wireless routing devices provided by many hardware vendors. A survey in 2013 showed that 71% still allow usage of TKIP, and 19% exclusively support TKIP.

WPS PIN Recovery

A more serious security flaw was revealed in December 2011 by Stefan Viehböck that affects wireless routers with the Wi-Fi Protected Setup (WPS) feature, regardless of which encryption method they use. Most recent models have this feature and enable it by default. Many consumer Wi-Fi device manufacturers had taken steps to eliminate the potential of weak passphrase choices by promoting alternative methods of automatically generating and distributing strong keys when users add a new wireless adapter or appliance to a network. These methods include pushing buttons on the devices or entering an 8-digit PIN.

The Wi-Fi Alliance standardized these methods as Wi-Fi Protected Setup; however the PIN feature as widely implemented introduced a major new security flaw. The flaw allows a remote attacker to recover the WPS PIN and, with it, the router's WPA/WPA2 password in a few hours. Users have been urged to turn off the WPS feature, although this may not be possible on some router models. Also note that the PIN is written on a label on most Wi-Fi routers with WPS, and cannot be changed if compromised.

MS-CHAPv20.25 in

Several weaknesses have been found in MS-CHAPv2, some of which severely reduce the complexity of brute-force attacks making them feasible with modern hardware. In 2012 the complexity of breaking MS-CHAPv2 was reduced to that of breaking a single DES key, work by Moxie Marlinspike and Marsh Ray. Moxie advised: "Enterprises who are depending on the mutual authentication properties of MS-CHAPv2 for connection to their WPA2 Radius servers should immediately start migrating to something else."

Hole196

Hole196 is a vulnerability in the WPA2 protocol that abuses the shared Group Temporal Key (GTK). It can be used to conduct man-in-the-middle and denial-of-service attacks. However, it assumes that the attacker is already authenticated against Access Point and thus in possession of the GTK.

Lack of Support of Forward Secrecy

WPA doesn't provide Forward secrecy, meaning that once pre shared key is compromised, an adverse person can decrypt all encrypted Wifi packets transmitted in the future and even past, which could be passively and silently collected by the attacker. This also means an attacker can silently capture and decrypt other's packets if a WPA-protected access point is provided free of charge at a public place like cafe because its password is usually shared to anyone with physical access without any special additional authorization. In other words, WPA only protects from attackers who don't have access to the password. Because of that, it's safer to use transport-layer security measure like TLS on top of that for the transfer of any sensitive data.

References

- Robert C. Newman (19 February 2009). Computer Security: Protecting Digital Resources. Jones & Bartlett Learning. ISBN 978-0-7637-5994-0. Retrieved 25 June 2010.
- Michael E. Whitman; Herbert J. Mattord (2009). Principles of Information Security. Cengage Learning EMEA. ISBN 978-1-4239-0177-8. Retrieved 25 June 2010.
- Tim Boyles (2010). CCNA Security Study Guide: Exam 640-553. John Wiley and Sons. p. 249. ISBN 978-0-470-52767-2. Retrieved 29 June 2010.
- Harold F. Tipton; Micki Krause (2007). Information Security Management Handbook. CRC Press. p. 1000. ISBN 978-1-4200-1358-0. Retrieved 29 June 2010.
- John R. Vacca (2010). Managing Information Security. Syngress. p. 137. ISBN 978-1-59749-533-2. Retrieved 29 June 2010.
- nitin.; Mattord, verma (2008). Principles of Information Security. Course Technology. pp. 290–301. ISBN 978-1-4239-0177-8.
- Anderson, Ross (2001). Security Engineering: A Guide to Building Dependable Distributed Systems. New York: John Wiley & Sons. pp. 387–388. ISBN 978-0-471-38922-4.
- Kohlenberg, Toby (Ed.), Alder, Raven, Carter, Dr. Everett F. (Skip), Jr., Esler, Joel., Foster, James C., Jonkman Marty, Raffael, and Poor, Mike, "Snort IDS and IPS Toolkit," Syngress, 2007, ISBN 978-1-59749-099-3
- Amoroso, Edward, "Intrusion Detection: An Introduction to Internet Surveillance, Correlation, Trace Back, Traps, and Response," Intrusion.Net Books, Sparta, New Jersey, 1999, ISBN 0-9666700-7-8
- "Exposing WPA2 security protocol vulnerabilities". Inderscience.metapress.com. International Journal of Information and Computer Security. 2014-03-13. Retrieved 2014-04-30.

Permissions

All chapters in this book are published with permission under the Creative Commons Attribution Share Alike License or equivalent. Every chapter published in this book has been scrutinized by our experts. Their significance has been extensively debated. The topics covered herein carry significant information for a comprehensive understanding. They may even be implemented as practical applications or may be referred to as a beginning point for further studies.

We would like to thank the editorial team for lending their expertise to make the book truly unique. They have played a crucial role in the development of this book. Without their invaluable contributions this book wouldn't have been possible. They have made vital efforts to compile up to date information on the varied aspects of this subject to make this book a valuable addition to the collection of many professionals and students.

This book was conceptualized with the vision of imparting up-to-date and integrated information in this field. To ensure the same, a matchless editorial board was set up. Every individual on the board went through rigorous rounds of assessment to prove their worth. After which they invested a large part of their time researching and compiling the most relevant data for our readers.

The editorial board has been involved in producing this book since its inception. They have spent rigorous hours researching and exploring the diverse topics which have resulted in the successful publishing of this book. They have passed on their knowledge of decades through this book. To expedite this challenging task, the publisher supported the team at every step. A small team of assistant editors was also appointed to further simplify the editing procedure and attain best results for the readers.

Apart from the editorial board, the designing team has also invested a significant amount of their time in understanding the subject and creating the most relevant covers. They scrutinized every image to scout for the most suitable representation of the subject and create an appropriate cover for the book.

The publishing team has been an ardent support to the editorial, designing and production team. Their endless efforts to recruit the best for this project, has resulted in the accomplishment of this book. They are a veteran in the field of academics and their pool of knowledge is as vast as their experience in printing. Their expertise and guidance has proved useful at every step. Their uncompromising quality standards have made this book an exceptional effort. Their encouragement from time to time has been an inspiration for everyone.

The publisher and the editorial board hope that this book will prove to be a valuable piece of knowledge for students, practitioners and scholars across the globe.

Index